Advance praise for *The Obesity Paradox*

"In this eye-opening book, cardiologist and medical researcher Dr. Carl Lavie shows that obesity has been unfairly blamed for a litany of health conditions, while its health benefits remain largely unknown. Reviewing scores of scientific studies that he and others have conducted, Lavie shows that while being heavier may put one at risk of developing certain diseases, carrying extra weight may be protective once illness has set in. Moreover, those who are thin but physically inactive face a plethora of unrecognized health risks. Challenging received wisdom with a careful review of medical research, this book is a must read for anyone who is trying to lose weight to improve their health or longevity."

—Abigail C. Saguy, associate professor and vice chair,
Department of Sociology, University of California, Los Angeles,
and author of *What's Wrong with Fat?*

"This provocative, phenomenal book by Chip Lavie will enhance the long-overdue paradigm correction, at a time when the obesity paradox is considered taboo by the mainstream opinion leaders."

—Kam Kalantar, MD, PhD, professor of medicine and pediatrics and
epidemiology, University of California, Los Angeles

"Chip is the rare triple threat: a great doctor, outstanding researcher, and gifted communicator. I have heard him present many times and read all his work on the obesity paradox, and I continue to find this topic absolutely fascinating. And no one understands it better than Chip."

—Timothy Church, MD, MPH, PhD, professor of preventative medicine,
Pennington Biomedical Research Center, Louisiana State University,
and coauthor of *Move Yourself*

"Paradox . . . revealed. Much more than casual social conversation, the debate around the 'obesity paradox' is critical to the health of nations. No one more than Dr. Chip Lavie has been a constant thoughtful observer on this topic—he has performed focused research and is an expert on the overall meaningfulness of body mass index and the obesity paradox. Indeed, Dr. Lavie introduces a whole new way of looking at body fat and its role in health. Once in your hands you will be wiser, better educated, and healthier!"

—Marc A. Silver, MD, FACP, FACC, FCCP, FAHA, clinical professor of medicine, University of Illinois at Chicago

"*The Obesity Paradox* is an important contribution to our understanding of one of the most interesting and unusual observations in medicine today. This book is highly recommended reading for all health-care workers and clinical investigators interested in the current global epidemic of obesity. Dr. Lavie has had extensive experience with the paradox in his research career, and his writing is clear and concise."

—Joe Alpert, MD, editor in chief of the *American Journal of Medicine*, professor of medicine, University of Arizona College of Medicine

"We are on the verge of disastrous and unprecedented health consequences as a result of our ever-evolving lifestyle choices. The solution to our problem is simple: above all else physical activity and fitness are essential to good health. Yet the human race remains lost, waiting for leaders to show us the way. Carl J. Lavie is one of those leaders. Dr. Lavie has extensively studied the importance of a physically active lifestyle. *The Obesity Paradox* is the culmination of his life's work and a shining beacon of knowledge, showing us the way to a healthier world."

—Ross Arena, PhD, PT, FAHA, professor and head, Department of Physical Therapy, College of Applied Health Sciences, University of Illinois at Chicago

"Dr. Chip Lavie is one of the leading medical researchers of cardiovascular illness, disease prevention, and the role of fitness in managing the multitude of risk factors associated with diminished health. *The Obesity Paradox* reframes the entire discussion of obesity and what it means to be healthy. Dr. Lavie focuses on the importance of fitness in making positive changes in the course of our health, despite being overweight or obese. To understand the big picture, before you try another weight-loss diet, read this book."

—Mark A. Williams, PhD, FAACVPR, FACSM, editor in chief of the *Journal of Cardiopulmonary and Rehabilitation Prevention*, professor of medicine and cardiology, Creighton University School of Medicine

"Carl J. Lavie, MD, is a world-renowned expert on obesity and the heart. In this book, Dr. Lavie describes the substantial body of evidence supporting the presence of an obesity paradox in cardiovascular and other disease states and then discusses the proposed mechanisms by which this phenomenon may occur. In some respects this book represents a contrarian view of the health effects of obesity, but one that is increasingly supported by high-quality research. This book will be of interest not only to professionals working in the field of obesity, but to anyone who wants to live a healthful life."

—Martin A. Alpert, MD, professor of medicine, University of Missouri–Columbia School of Medicine

"Dr. Chip Lavie is an outstanding clinical cardiologist who has reached a level of achievement matched by very few scientists. One of the world's leading authorities on the topic of the obesity paradox, he has now used his extensive scientific background and clinical experience to write a book for the public. I am very excited that he has written a book on this crucial issue. We need to focus much more on healthy lifestyles, and in particular a fit and active way of life. For health, longevity, and function, your activity habits and physical fitness levels are far more important than your body weight. Read this book and focus on what is important."

—Steven N. Blair, PED, professor, Departments of Exercise Science and Epidemiology & Biostatistics, University of South Carolina

"In this wonderful book, Dr. Lavie displays the extraordinary talent of articulating complex concepts regarding body size, metabolism, and wellbeing in a manner that is sophisticated and precise, but also remarkably straightforward and clear. He spells out issues of obesity vs. slimness that affect almost everyone, clarifying relative benefits and risks of different body shapes, as well as the related impact of exercise, age, disease, and other key dynamics on overall health."

—Daniel Forman, MD, associate physician, Brigham and Women's Hospital, associate professor of medicine, Harvard Medical School

"Finally, a book that sets the record straight about the leading health problem facing our nation: obesity. As one of the world's leading experts on preventive medicine, Dr. Carl Lavie brings to bear the latest scientific research on what it means to be truly 'healthy.' *The Obesity Paradox* is a product of years of collaborative research on the myths of body fat, calling out that health is more than just a number on a scale. *The Obesity Paradox* is the best book I've read on the roles of fatness and fitness, and is a must read for anyone interested in good health."

—Richard Milani, MD, MD, FACC, FAHA, chief clinical transformation officer, Ochsner Health System, vice chairman, Department of Cardiology

"Bad fat, good fat. In this book, Dr. Lavie uniquely melds science and art into a fantastic exploration of heart health that shines a bright light on the most important epidemic of our times."

—Mandeep R. Mehra, MD, FACC, FACP, medical director, Brigham and Women's Hospital Heart and Vascular Center Executive Director, Center for Advanced Heart Disease, professor of medicine, Harvard Medical School

"Dr. Lavie has devoted most of his research life to unraveling different facts of obesity—a worldwide health problem. He is one of the leading experts on the so-called obesity paradox, the idea that fat can protect you from an early death. And in this book the reader will be able to understand the clinical issues associated not only to obesity but the obesity paradox. It will change the reader's views about fatness, offering a valuable new perspective."

—Hector Ventura, MD, FACC, FACP, director, Cardiovascular Disease Training Program and Education at the Ochsner Clinic Foundation, and professor of medicine, Tulane University School of Medicine

"Two out of three Americans are overweight or obese, and many of them feel guilty and/or fearful about these extra pounds. Dr. Lavie, a brilliant and internationally renowned cardiologist who is one of the foremost experts on preventive cardiology, explores the long-term health consequences of carrying around excess fat, offering startling and counterintuitive findings. This groundbreaking book is a must-read for anyone who is overweight or obese and struggling to lose weight, often times through using potentially dangerous therapies including radical gastric bypass, dangerous drugs and/or supplements, and crash dieting leading to detrimental yo-yo weight changes. Dr. Lavie's insightful book has the power to change the zeitgeist about body fat—and what a truly healthy shape looks like."

—James O'Keefe, MD, medical director of the Charles and Barbara Duboc Cardio Health and Wellness Center at Saint Luke's Mid America Heart Institute, and professor of medicine, University of Missouri-Kansas City

"I was initially skeptical: Did we really need another book on our overweight/obesity epidemic? But then I started reading *The Obesity Paradox* and I could not put it down: It is well-written, doesn't rely on medical jargon, and draws from the most up-to-date and accurate medical literature. Dr. Chip Lavie, a 'giant' in the fields of preventive cardiology/lifestyle medicine/cardiac rehabilitation, provides the reader with a research-based hopeful message: It's not doomsday to have some extra fat on you, especially if you maintain a certain level of aerobic fitness. If you choose to read only one health-related book this year, read this one. I plan to recommend it to all of our patients."

—Barry A. Franklin, PhD, director, Preventive Cardiology and Cardiac Rehabilitation William Beaumont Hospital, and professor of internal medicine, Oakland University William Beaumont School of Medicine

The OBESITY PARADOX

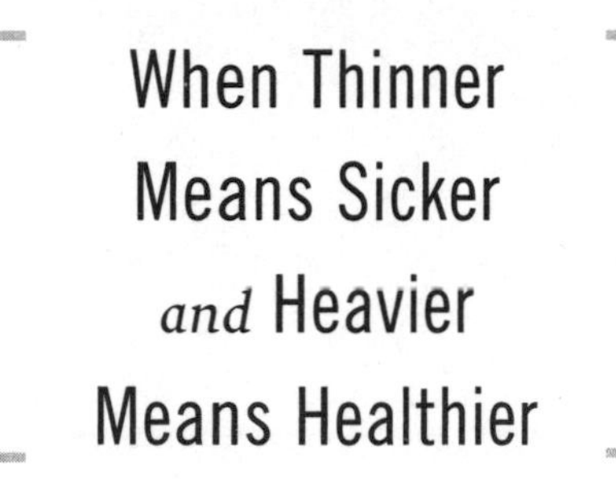

CARL J. LAVIE, M.D.

with Kristin Loberg

HUDSON STREET PRESS
Published by the Penguin Group
Penguin Group (USA) LLC
375 Hudson Street
New York, New York 10014

USA | Canada | UK | Ireland | Australia | New Zealand | India | South Africa | China
penguin.com
A Penguin Random House Company

First published by Hudson Street Press, a member of Penguin Group (USA) LLC, 2014

LIBRARY OF CONGRESS CATALOGING-IN-PUBLICATION DATA
Lavie, Carl J., author.
The obesity paradox : when thinner means sicker and heavier means healthier / Carl J. Lavie, M.D. ; with Kristin Loberg.
pages cm
Includes bibliographical references and index.
ISBN 978-1-59463-244-0
1. Obesity—Health aspects. 2. Overweight persons—Health and hygiene. 3. Human body—Composition. 4. Weight loss—Physiological aspects. I. Loberg, Kristin, author. II. Title.
RA645.O23L38 2014
616.3'98—dc23 2013045483

Printed in the United States of America
10 9 8 7 6 5 4 3 2 1

Set in New Caledonia
Designed by Eve L. Kirch

To the pursuers of health and happiness . . . no matter how much you weigh.

And to anyone who has felt tyrannized by matters of weight, dieted to no avail, and searched for the secret to longevity.

Paradox: a statement that is seemingly contradictory or opposed to common sense and yet is perhaps true.

Contents

Introduction: The Miseducation of Fat

Take a moment to consider what the word *fat* brings to mind. This small, three-letter word looms large in our collective consciousness today, arguably as powerful in our vocabulary as the word *no*. And indeed, these two words seemingly share the same negative vibe. For many, *fat* conjures vivid images of unhealthiness, greasy foodstuffs, fleshy folds, and even ugliness, slovenliness, and gluttony. We rarely see fat individuals gracing the covers of popular magazines. In fact, we see the absolute opposite, as if extreme thinness were the definition of beauty and health.

I don't think any one of us can get through the day without this word creeping into our minds, however subliminally. We are bombarded with reminders that we live in a society in which more than two-thirds of us are overweight or obese. We are told to lose weight everywhere we look, from the packaging of food products that market their brands based on low calories, weight loss promises, and other health claims, to the phalanx of media stories, books, and articles that relentlessly cover the obesity epidemic and try to show us how to trim down. We know it's a problem. Most everyone also has a clear idea about what excess fat does to harm the body. It wreaks havoc on our metabolism, opens the door to diabetes and heart disease, among other dangerous health conditions, and basically increases our risk of dying young.

But here's a question that we must ask ourselves: Have we demonized fat to our own detriment? Much in the way a glass of wine a day has been proven to impart health benefits, but four or five glasses puts us at risk for an untold number of health challenges, can body fat in the right amount (and more than what you're probably currently striving for) be exactly what we need to live long and enjoy the highest quality of life possible? Put simply, has our modern culture duped us into thinking excess body fat is bad no matter what and should be burned away at all costs?

I'm here to tell you that fatness has been sorely misunderstood and wrongly portrayed in many respects. By the same token, endurance sports and rigorous exercise have been unjustly glorified. Similar to how the famous French paradox points to an inverse relationship between the incidence of coronary heart disease and the consumption of saturated fats, the so-called "obesity paradox" points to an inverse relationship between body fat and risk of death in many cases. While it's well documented that obesity has contributed to our challenges with chronic illness, and that exercise is often a surefire antidote, no one has explained some of the stranger cause-and-effect reactions going on deep within our cellular makeup that have everything to do with fat's positive side. It's pretty straightforward: Fat isn't always bad. And exercise isn't always good.

These are hardly inconsequential findings: The fact that body fat protects us in a lot of surprising ways—not to mention may help us live long after a grim diagnosis or heart attack—has been quietly circulating in our most prestigious medical literature for the past few years. And so has the certainty that too much exercise can shorten our life span. I have led the charge in some of these findings through my work as a cardiologist (and avid runner) with a keen interest in the relationships among the subjects of the heart, fat, obesity, exercise, disease, and longevity. The information that my research, alongside that of many others, has amassed in this realm is truly staggering and begs to be known by the public, which is increasingly deceived by an anti-fat, pro-exercise industry. It has also led doctors and scientists like me to question so much about what we consider to be "healthy."

When I asked you to think about the word *fat* at the beginning of the introduction, did anything positive and beautiful come to mind? Now, wouldn't it be great to be able to relax more about issues around weight? Don't get me wrong: I won't endorse a fattening lifestyle, whereby I recommend avoiding treadmills and embracing daily cinnamon buns. I'm not promoting fatness, and I won't be suggesting that people of "normal" weight start intentionally packing on the pounds. The message is a bit subtler than that, but I trust you'll find it equally as powerful and liberating. And if I had to sum up the main message of this book, it would be this: Contrary to what the popular media broadcasts and what the general medical wisdom has you believing, it's not doomsday to have extra fat on you, especially if you maintain a certain level of fitness (and I'll explain exactly what "fitness" means, for it's not being able to run a six-minute mile or commit to an exercise regimen on a par with an athlete's). For millions of people categorized as "overweight" or "mildly obese" by the most commonly used standard today—the body mass index, or BMI—the achievement of optimal health may mean staying exactly where you are in terms of weight. That's right: You don't have to lose weight and you don't have to set your sights on getting your BMI down to less than 25. You may, in fact, be much better off sustaining a BMI of between 25 and 30, or even slightly above that, depending on your body type. I will show you why and how this is possible. And if you do choose to lose weight, make sure you're doing it for the right reasons, a topic we'll also discuss.

As a practicing cardiologist who cares day in and day out for individuals struggling to cope with serious heart trouble, I'm compelled to get to the bottom of all this. Perhaps it's because I'm not just a board-certified cardiologist at one of our nation's most respected heart centers (John Ochsner Heart and Vascular Institute, in New Orleans, Louisiana, a top-twenty nationally ranked cardiac program); I'm also the author or coauthor of more than eight hundred medical publications, including two cardiology textbooks and almost forty book chapters.

But, as I've hinted at, very little of this illuminating data recorded in the halls of medicine has landed in books for people like you—people who want to know the "secrets" to a vibrant, long life, some of which might go against conventional wisdom. And, as you're about to find out, the new science most certainly does compete with what you've been taught to believe. My clinical roles and work in research give me a unique perspective. Add to that my passionate drive to run more than I probably should, and you've got someone who will go to the moon and back to decode the riddles of superior health. And let's not leave out the fact I live and work in the most obese state in the nation here in Louisiana.

As a scientist who must always obey the rigors of the scientific method, shun biases, and not turn a blind eye to any possible influencers and feasible variables, I will leave nothing to imagination or conjecture as we look at all the puzzle pieces individually before putting them together to draw some valid conclusions. This will entail an engrossing, fresh look at some basic physiology, the definitions of body mass index and fitness, the interplay of genetics and the environment, and the way science typically handles a sudden change in thought, perspective, or dogma.

Have Your Cake and Eat It Too

I've organized this book into three parts. In "Part I: A Paradox for the Ages," I'll take you on a tour of obesity—what it really means, how it negatively impacts the body, how fitness can trump fatness (spoiler alert: it's the key ingredient in health *in spite of excess weight*), and how science explains the obesity paradox. This is where I'll describe why being thin and unfit, as opposed to overweight or moderately obese and fit, is probably the most harmful body type to have if you want to live happily to a ripe old age. I'll show you how body fat can be life sustaining and why we may want to think twice about combating the inexorable weight gain we experience as we age. I'll also use Part I to show the scientific proof of how you can be fat, fit, and remarkably healthy. In "Part II: The Purpose and Power of Fat," I'll offer an in-depth exploration of the facts about fat. Why do we need it? What does it do to sustain life? I'll present the latest research and explain the rationale for people who appear slim and trim yet suffer from metabolic conditions like type 2 diabetes—a so-called "lean paradox."

Indeed, looks can be deceiving. You'll come away from this conversation with a clear understanding that the real killer today isn't necessarily obesity. What's much more damaging to our health and well-being is not only the immense pressure we put on ourselves to lose weight and follow unrealistic diets, but the metabolic harm we inflict on our bodies long before obesity becomes established. This metabolic injury often results from a lethal combination of chronic blood sugar imbalances and inactivity—both of which have absolutely nothing to do with weight per se and can occur in someone who looks slender and seemingly epitomizes health on the outside.

By dispelling myths about facts we've all come to accept as doctrine, such as the importance of BMI and waist circumference, I hope to have you questioning your beliefs and rethinking core assumptions about fat and body weight. The research is so captivating that it will draw those on both ends of the spectrum—thin, sedentary people who regulate their weight through diet alone, and the morbidly obese who haven't exercised in years. It will also speak to everyone in between who encompasses a mix of individuals at various weights and fitness levels, some of whom take fitness to an extreme.

Finally, in "Part III: Striking the Balance," I'll wrap up the book with a look at how anyone can take advantage of this new knowledge. This is where I'll help you assess where you stand in the scope of all the risk factors already detailed, and I'll present some general guidelines on how you can "have your cake and eat it too." This includes fundamental lessons on diet and exercise with simple steps to take and strategies to consider. I'll also use this part to address many frequently asked questions, such as: Which dietary protocol is best? Which fitness routine is ideal? When does exercise become dangerous and health depleting? Should the goal for everyone be to prioritize weight management or to prioritize fitness level? Are there any drugs or supplements that one should consider taking to manage health and weight? I'll answer these questions for both those who are hoping to avoid a chronic illness and those who have been diagnosed with one. To be clear, this book is for people who are already living with a chronic illness, as well as those who hope to prevent one. It's as much for young, skinny individuals as it is for older folks struggling with those stubborn extra pounds. And it speaks to active people who think they are fit as well as those who prefer a more sedentary lifestyle.

Every day we hear some new piece of news related to health. We are bombarded by messages about our health—good, bad, and confusingly contradictory. And we simply accept most of these messages. Even the smart, educated, health-conscious, well-read, cautious, and skeptical people are quickly convinced. It's hard to separate truth from fiction. It's hard to know the difference between what's healthful and hurtful when the information and endorsements come from apparently the brightest, most impartial of people.

If you consider some of the advice doled out in the past 150 years from so-called experts, you'll quickly realize that many things do not always appear as they seem, and that it's quite common to witness a complete about-face when it comes to the validity of a certain fact, claim, or practice. At one time or another, bloodletting was thought to cure illnesses, baby formula was recommended over breast milk, and doctors endorsed cigarette smoking. Only recently, believe it or not, have we really begun to understand how the choices we make about what to eat play into health and longevity.

When I imagine the world fifty or so years from now, I wonder what kind of bogus claims that we accept at face value today will have turned out to be totally false. I also wonder, given the research my colleagues and I have been collecting, if the conversation on fat will have shifted. Can fat, one of the most fiercely debated topics of our time, gain a better reputation in the future?

So, please, stop ruminating over your current weight and guessing what you must lose. In addition to gaining a lot of useful information in this book, you also might come to the conclusion that you're closer to the model of ideal health than you think.

{ **PART I** }

A Paradox for the Ages

{ CHAPTER 1 }

Obesity: What Does That Really Mean?

Everyone has a clear picture of what obesity looks like. The word originated in the seventeenth century from Latin: *obesus*, which literally means "having eaten until fat." *The Oxford English Dictionary* documents its first usage in 1611 by Randle Cotgrave, an English lexicographer who compiled a bilingual French and English dictionary. We may think that obesity is a relatively new disease that emerged in the late twentieth century, but it dates back to the prehistoric era. Twenty-five to thirty-five thousand years ago, the first sculptural representations of the human body depicted obese females. At various times over the centuries it has stood for wealth or stature, or even luck, in the case of plump Venus figurines used in primeval rituals. When it was deemed good and a sign of health, it typically signaled that someone was living a privileged life and didn't go hungry. That person was also likely to be just moderately overweight or mildly obese—not morbidly so, as we see among 3 percent of the US population today. And at other times throughout history, obesity has been looked upon as a negative attribute, a character flaw, and an insult to health. In the sixth century BC, the Indian surgeon Sushrutha related obesity to diabetes and heart disorders. Sushrutha is actually considered the father of plastic surgery today, made famous for not only his nasal reconstructions at a time when medicine was in its infancy, but also for his futuristic understanding of human anatomy, physiology, disease, and relevant therapies.

Virtually every culture on the planet has recorded this age-old condition, from the ancient Egyptians and ancient Chinese to the Aztecs and Hippocrates, the father of modern medicine, who wrote around 450 BC (about 150 years after Sushrutha) that "corpulence is not only a disease itself, but the harbinger of others." These formal writings credited the Greeks with first recognizing obesity as a medical disorder.

Only in the past forty years has obesity become much more prevalent . . . and much more demonized by society. Well over 70 percent of us are now overweight or obese (compared with less than 25 percent some forty years ago); less than a third of us are considered at a "healthy weight." But what, exactly, does that mean? What constitutes a healthy weight? In fact, what constitutes the word *weight*? Can having a lot of muscle mass and, therefore, a high body mass index (BMI) necessarily equate with "obesity"?

Before we can even begin to address such questions, it helps to understand that all these terms—*obesity*, *health*, *weight*—are very loaded words and perhaps always have been. How we define each of these depends a lot on context. Contrary to what you might think, the definition of obesity, or just being overweight, remains controversial. In the US, mortality data provided by the Metropolitan Life Insurance Company (MetLife) historically has been used to define obesity. In fact, the entire concept that excess weight can make for an early grave first came from studies done by our country's insurance industry, not the medical community. Insurance companies may not be in the business of healing, but they are perfectly suited for the task of figuring out who is likely to die sooner rather than later. Yet their data related to mortality only—the chance that someone would die based on his or her body weight. This definition depends solely on a person's frame (i.e., size, as defined by weight and height). You might recall seeing versions of the charts created by MetLife and prominently displayed in doctors' offices when you were growing up. The chart listed the difference between small, medium, and large body frames based on height in inches and weight in pounds. Its purpose was to give ideal body weight values and provide an estimate of body fat composition. But how we came to draw the lines among "small," "medium," and "large" (and, for that matter, normal versus obese, since only height and weight measurements were considered) was quite arbitrary. The original definition was not connected to or based on obesity-related diseases or death.

Today we equate obesity with epidemic, and as such we define and conceptualize it differently than ever before. Indeed, the obesity epidemic as we give meaning to it in the twenty-first century is not particularly old. Although obesity in adults and children, male and female, has doubled over the past forty years, the biggest increase has been since 1980. Prior to this, evidence of the obesity epidemic was hardly identifiable. Obesity rates among children born between the 1930s and most of the 1970s, for example, remained low and steady. And then the numbers started to go upward rapidly through the 1980s and 1990s.

To get a sense of what was happening among adults before national surveys began in the 1960s, we can turn to government records of measurements taken on young men who were drafted into the military. From those records, we know that, on average, men grew about 1.5 BMI points heavier in the hundred years following the Civil War. That translates to about ten pounds more for a man of average modern height, which is indeed a meaningful change, but nowhere near as significant as the shift that took place between 1980 and 2000. During that almost-twenty-year period, the average man's BMI went up by 2.3 points, or roughly sixteen pounds. Another way of looking at these numbers is to say that in a matter of two decades, the average BMIs of young men climbed seven times faster than the rate calculated over the previous hundred years. So while it may be true that our average weight as measured by BMI has been slowly but steadily rising for a long time, in recent decades we've witnessed a spiraling surge that has driven many adults and children into the bona fide obesity category. Meanwhile, some data has suggested that we've also experienced a wild leap in caloric intake.

According to the US Department of Agriculture, the average American daily caloric intake increased massively within a matter of just two generations. In 1970, we ate closer to 2,234 calories a day; this amount exploded to 2,757 by 2003. And, obviously, eating more than 500 extra calories each day can lead to significant weight gain, if you're not out there pounding the pavement in a vigorous physical activity to burn off the extra energy. Of those 523 additional calories, 292 have been attributed to added fats, oils, sugars, and sweeteners while more grains, many of which are refined, accounted for the rest (188 calories).

Between 1977 and 1995, our consumption of fast-food meals tripled and caloric intake from those meals quadrupled. By 1997, our obesity rate had gone up to 19.4 percent (from a 13 percent rate in 1962). That same year, the World Health Organization (WHO) held its first meeting on the subject, in Geneva, Switzerland, which resulted in the introduction of new criteria for "normal weight" (BMI of 18.5–24.9), "overweight" (BMI of 25–29.9) and "obese" (BMI of 30 or higher). And not even a year later, in 1998, the CDC lowered its BMI cutoffs to match the WHO's classifications. Ten years after that, in 2008, US obesity rates appeared to be rising indefinitely, reaching 33.8 percent; worldwide, the WHO claimed that 1.5 billion adults 20 and older were overweight, and of these more than 200 million men and nearly 300 million women were obese. By now, people were beginning to understand the hotly contested case against sugar, especially refined processed sugars like high fructose corn syrup, as being a prime suspect in the epidemic (as opposed to just eating more fat). Sugar also helped explain why obesity was becoming a problem not just in high-income countries but also in low- and middle-income countries, particularly in urban settings where cheap calories were leading to overfed, undernourished individuals.

Since the dawn of recorded history, famine and malnutrition have been the scourge of humankind—that is, until relatively recently (and especially so in developed nations). The current obesity epidemic is now considered a public health crisis globally. But this could not have happened prior to the technological advances of the eighteenth century, which allowed for our food supply to become so grand and reliable. These advances had an immediate and positive effect on our health and longevity; improvements in the quality, amount, and variety of our food led to a boost in our body size and, in turn, health. We got thicker and lived longer. But too much of a good thing turned ugly, for since the Second World War we've experienced an overabundance of too-easy-to-get food. At the same time, we've lost the need and incentive to engage in physical activity, largely due to technologies that have benefited us in other ways. Combined, these two forces—more food, less exercise—rendered a trend toward obesity.

Which begs the question: What is the more powerful factor in causing obesity—more food or less exercise? As I've implied, some evidence has pointed to increased caloric consumption, especially from sugar and fast food, as the principle cause of obesity. But my colleagues and I have published studies suggesting that the rise in caloric intake has been much less appreciable than most people realize, and that the real culprit is energy expenditure: The number of calories burned daily has declined enormously during the past five decades, far outpacing the increase in our daily bread (and sugar). And this, in my opinion, is much more related to the epic increases in weight noted over time than our voracious appetite. Many of my colleagues agree; Dr. Timothy Church, from Pennington Biomedical Research Center, in Baton Rouge, demonstrated this very fact in a 2011 paper, suggesting that the dramatic drop in the average amount of energy expended in work-related activities over the past fifty years almost totally explains the increasing prevalence of obesity during the same time period.

In February 2013 I coauthored a study with Drs. Ed Archer and Steven Blair, from the University of South Carolina, that examined trends in women's use of time and household energy expenditure over the past forty-five years. Although my colleagues and I took some flak from the popular press for focusing squarely on women performing labor-intensive household chores like vacuuming, cleaning, and cooking, the results—a staggering over eighteen hundred fewer calories burned on a weekly basis—pointed to broader trends that could be interpreted widely and include men. In the December 2013 issue of *Mayo Clinic Proceedings*, we focused on the marked decline in energy expenditure in mothers during the past fifty years, which is even more dramatic in those with very young children. Such a trend not only plays a major role in the obesity epidemic among mothers, but it also may impact the physical activity patterns and prevalence of obesity in their children (or the next generation).

We all can concede that we do less today from a physical standpoint than we did half a century ago. We're more likely to delegate household chores to hired hands and resort to eating out rather than cooking (and cleaning up afterward). We work longer hours, usually at a desk, and spend more time in front of electronic devices and the TV than playing outside or making an effort to engage in formal exercise. We mow our lawns with gas-powered machines rather than old-fashioned push mowers, drive to local destinations, favor the elevator or escalator over stairs, and find many conveniences at the touch of a button rather than requiring any physical movement. So contrary to popular belief, our thicker waistlines are not just the outcome of calorie overload, even when those calories come from traditional American fast-food fare, such as burgers, fries, doughnuts, and sugary beverages. While it's true that fewer calories burned means caloric consumption must be reduced to prevent excess weight gain, overweightness, and obesity, the real cause of obesity is not excess caloric consumption but rather physical inactivity.

Regardless of the cause, the numbers in 2010 ticked even higher, as obesity rates reached "catastrophic levels" at 35.7 percent of adults and approximately 17 percent (or 12.5 million) of children and adolescents aged 2 to 19 years. Study after study highlighted the costs associated with such a statistic; one 2010 study concluded that medical expenses directly related to obesity are costing America 160 billion dollars per year. And the estimated indirect costs devour a total of 450 billion dollars annually. Being obese or overweight poses a major risk for chronic diseases, including type 2 diabetes, cardiovascular disease, hypertension and stroke, and certain forms of cancer. In fact, it's estimated that every third person born in 2000 will have type 2 diabetes as an adult. According to the American Heart Association, 70 percent of diagnosed heart disease cases are linked directly to obesity. More than half of adults with type 2 diabetes nationwide are obese, and 30 percent or more are overweight. It's been projected that by 2030, almost half of Americans (41 percent) will be obese, while a whopping 86 percent of us will be overweight.

Of course, researchers have pegged other costs from their perch on the antiobesity crusade, coming up with figures like how, per year, obesity costs the average man an extra $2,646 and the average woman an extra $4,879. According to some experts, obese people are paid, on average, about 2.5 percent less than their thinner counterparts; other findings show that obese women are paid 6 percent less than their thinner peers doing the same work. As of 2012, we've led the charge globally: Of all countries, the US has the highest rate of obesity.

Suffice it to say obesity as a chronic disease with well-defined health consequences is less than a century old. Only in the past year has the US come to categorize obesity as a "disease" from a medical standpoint, though this was partly a political move. Nevertheless, those in the medical community have recognized obesity as a disease to some degree or another for a long time.

Obesity is a fascinating and destructive phenomenon when viewed within the context of its costs to health from both an individual and a societal perspective. But again, one distinction we need to make is this: Where is the line between fat as a "good" thing and fat as a "bad" thing? Aside from all these eye-blurring numbers, how can we accurately define obesity as a true disease when we know that people can be in fact healthy but overweight or mildly obese at the same time? Obesity is intriguing in that it's uniquely human or human caused. Have you ever seen an obese wild animal? Obesity doesn't exist in nature because slow, fat creatures are quickly killed by predators, or they die of starvation if they are inept predators. But the term *obesity* is human conceived. We can define it however we want, and have been vague at best in finding an accurate definition that takes into consideration all the conflicting data emerging from an intractable medical community. What kind of clashing, contradictory data? Let's go there next, and then circle back to all these numbers, statistics, and definitions.

A Big Fat Conundrum

It was more than a decade ago that I began to document the obesity paradox in heart patients. I'll save the in-depth version of the story for chapter 5 but give you a quick primer here. What doctors like me were noticing was that those who were overweight or moderately obese and had a form of cardiovascular disease often experienced a much better prognosis than their thinner counterparts. In other words, they lived *longer* after being diagnosed with an illness that was likely brought on by their weight and potentially fatal in the short term. (To my knowledge, the expression *obesity paradox* was first coined by Dr. Luis Gruberg and colleagues in 2002 at the Cardiac Catheterization Laboratory and the Cardiovascular Research Institute in Washington, D.C. Contrary to their original hypothesis, they found that overweight or obese patients, as opposed to thinner patients, had roughly half the risk of dying within a year after undergoing angioplasty, a procedure to treat narrowed arteries in a diseased heart.)

So, as you can imagine, this didn't seem to make much sense. It was grossly counterintuitive and illogical. I kept asking myself, If obesity causes heart disease, then how can it suddenly turn protective once one has such a malady? What had we miscalculated or misunderstood when it came to fat?

To say this discovery ruffled a few feathers in my field is an understatement. As I probed further, I suddenly found myself swimming upstream against a tidal wave of ingrained ideology and unwavering belief systems. No one was prepared to entertain new thinking about fat, much less a perspective that put a radical spin on established wisdom about excess fat. Even veteran scientists and respected journal reviewers whom I assumed to be objective were skeptical upon looking at the data. I knew, however, that the science would eventually speak for itself. And it did. Over the past decade a multitude of highly respected studies from around the globe have confirmed the obesity paradox, and the reluctant scientific community has been forced to rethink its definitions of fat—and consider how it could actually have positive attributes under certain circumstances. New research has also alerted the medical community to the serious hazards of having a body type so coveted by the mainstream public and modern beauty industry: superthin. Willowy, skeletal bodies grace covers of magazines, not plump full-figured individuals (many of whom can probably outrun and outlive the runway models).

However, perhaps the most compelling observations that give credence to the undeniable data have absolutely nothing to do with heart disease at all and everything to do with other conditions. This happens frequently in medicine: What's true for one condition is often true for others with factors in common. As it turns out, the obesity paradox has been documented in a host of chronic ailments in addition to those related to the heart. And the laundry list is indeed a mixed bag of very different health challenges that share a most astonishing contradiction in character when a surfeit of fat is around. These include afflictions like arthritis, kidney disease, diabetes, cancer, and even HIV infection. We often attribute excess weight to an increased likelihood that these conditions will be worsened or aggravated as a result, but the evidence proves otherwise: People who have been diagnosed with any of these ailments fare better in the long run if they are overweight or even mildly obese than if they are normal weight. This fact remains true even when researchers rule out weight loss attributable to other preexisting illnesses, such as cancer, that could make thinner groups appear less healthy.

What explains this paradox?

The Obesity Paradox: Overweight and moderately obese patients with certain chronic diseases, from heart disease and arthritis to advanced cancer and even AIDS, often live *longer* and fare better than normal-weight patients with the same ailments. Case in point: Patients A, B, C, and D are all 54 years old and each has just suffered a heart attack. Patient A has a BMI of 23 (normal weight); Patient B has a BMI of 27 (overweight); Patient C has a BMI of 30 ("obese"); and Patient D has a BMI of 18 (underweight). It turns out that Patients A and D are much more likely to die in the next several years for a host of potential reasons we'll be exploring (and between the two, Patient D—the very thin individual who is likely lionized by society for his or her figure—is probably particularly doomed). It's a cruel irony that we've only recently discovered and that goes against so many entrenched assumptions and stereotypes. And it's one that the scientific literature fought to acknowledge for far too long. But over the past decade, this finding has been proven by my studies and those of other prominent scientists around the world. Now the scientific community has had to reexamine its definition of fat and consider how it could actually be a good thing.

Hence the whole purpose of this book. We're going to tackle this very conundrum at length, which will be an enlightening exercise in scientific analysis as much as it is an exercise in thought. I will teach you how to think through wildly divergent conclusions, how to decipher fact from fiction, weigh all the potential *confounding factors* (additional, often hidden variables or factors that can skew results in a given study—in the obesity paradox research, such confounding factors include access to health care and lifestyle habits such as smoking), and how to gain an appreciation for the ways in which a scientist is obliged to examine results, consider inconvenient contradictions, and forge new investigations in the quest for the truth. Because that's what we're really dealing with here: inconvenient research, which, sadly, rarely—if ever—finds its way to consumers. (Another question to ask: If this is the very first time you've heard of such a paradox, do you wonder why? As part of our discussion, I'll disclose to you some of the politics involved with making headlines today and how some of the most important knowledge frequently remains under wraps as if it were classified information.) The obesity paradox is just too significant to stay buried in the medical literature. Its implications are far too grave for millions of people—especially those who struggle mightily with their weight. If the science gives us permission to enjoy a few extra pounds and padding for the sake of a better, longer life, shouldn't we be listening?

The Power of a Paradox

Prior to entertaining all the possible reasons for the paradox from a scientific perspective, we have to start with the more fundamental questions: Does having a bigger waist circumference increase your odds of living longer and actually *surviving* a heart attack, cancer, kidney disease, and deadly infections? Have we accelerated our decline as we age by pushing low-fat body types at the expense of muscle mass and healthy body fat? Can obesity, dare I suggest, be a blessing to some extent? When we fight natural weight gain as we age, are we doing a great disservice to ourselves? Could there be a genetic, evolutionary reason to be a little chubby when we're older and at greater risk for disease (and not beat ourselves up while trying to get down to what we think is an ideal weight)? Much like having increasingly thinner, grayer hair and more wrinkles as we grow older, are those extra pounds we carry in older age a reflection of just that—age?

These were among the questions that started to run in my head when I first came across the mystery. And as I started to probe deeper, I simultaneously was pressed to look at more than just the fat component. I had to consider how fitness factored in, because over and over again I noticed that the people who fared the worst weren't just low in body fat, but they were low in muscle mass and cardio fitness, too. So it's not only the presence of fat that helps us to live long past a horrible diagnosis of a chronic condition, but it's also the existence of what's called *cardiorespiratory fitness*. There's a great divide between being just fat and being fit and fat, as we'll scrutinize in rich detail.

Of course, delving into the fitness side of the story will also bring up a lot of other interesting facts—and enigmatic inconsistencies—that science has recently culled from years of research. As the obesity paradox has come into clear view over the past decade, so have the facts surrounding the limitations of exercise. We all know that exercise is good for us, lowering our risk for just about everything, but how much is too much? Does running long distance or working out for more than an hour a day weaken our hearts and shave years off our lives? When does the law of diminishing returns come into play? Sooner than we think, as I'll explore with you. No doubt the debate in the fitness community has been equally as heated and contentious as the one around the obesity paradox—just try telling an elite triathlete or die-hard marathoner to consider a new sport.

Although my good friend and colleague, Dr. James O'Keefe, and I have been criticized for cautioning people about the perils of extreme exercise, the verdict will be unanimous and obvious once you know the science: The unique and almighty benefits of exercise are best gained by moderate physical activity. The kind of exercise we need to do to win marathons and condition our bodies to be those of elite athletes are very different from those that support cardiovascular health and overall longevity. As my colleagues and I have stated numerous times in lectures and published studies, "We are not so much born to run as born to walk." And, for the record, I'm amused but saddened by the criticism. My mission is to help people prevent and avoid heart disease, which remains our number one killer. When the science informs that agenda, I stick with it. While it's true that a major public health problem for our society as a whole is too little exercise, the most important public health concern about overdoing exercise is that it may erase the potent cardio-protective effects of long-term moderate-dose exercise. Note the parallels: Just as too much or too little fat is immensely destructive to health, so is too much or too little exercise.

The scientific relationship between our body's fat and our risk for illness and even death is not well understood by most people, including doctors who were educated years before this new science was established. It's time we paid attention to the latest research, which demands we change long-held convictions about fat. After all, the statistics pointing to this obesity paradox across a spectrum of disorders are astounding:

- Diabetes patients of normal weight are *twice as likely to die* as those who are overweight or obese.
- Heavier dialysis patients have a *lower chance of dying* than those who are of normal weight or underweight.
- Mild to moderate obesity poses *no additional mortality risks* to those already suffering from heart disease.
- Being overweight is *not* related to increased mortality in the elderly.
- Obesity can help someone with cancer or an infection such as HIV live longer.

A good paradox in science is a good problem to have if you're looking for the truth; it opens the door to new information.

What It Means to Be Healthy

So if you were to ask me what it means to be healthy, my response would be the following: Regardless of your weight (and BMI), if you're metabolically sound—you don't have serious metabolic conditions like uncontrolled type 2 diabetes or lipid disorders such as dangerously high cholesterol or significantly elevated blood pressure—and you are cardiovascularly fit, you are healthy. This means you can be considered overweight or even mildly obese by modern standards but still healthy by many other measures. And the higher your fitness level, the better. I will even go so far as to suggest that you can be overweight or mildly obese with some metabolic disorders as a result and still be deemed healthy. You can be *healthier* than someone with a lower BMI, for reasons I'll describe (hint: The fitter you are, the less your weight or BMI matters).

Don't get the wrong idea when I stress the importance of fitness and your mind immediately turns to an over-the-top, unachievable image of what that means. You needn't become a gym rat or contender for the Olympics to be considered "fit" and, in turn, healthy. I will clearly outline what defines "fit" and show how you can increase your level of fitness without feeling like you're forcing yourself to go to any extremes or pretend to enjoy sports.

One of the chief debates I'll address throughout the book is the value of body mass index. And I will encourage you to rethink the meaning of the term *BMI* in your personal definition of health. While it continues to be a benchmark of sorts for classifying weight categories, it should not be used strictly to define "healthy" versus "not healthy." Published in 1996, BMI replaced old charts of ideal weights for various heights. BMI measures a person's weight (in kilograms) divided by the square of his or her height (in meters), and is believed to be a more precise measure of body fat based on height and weight. As noted previously, in the US we currently follow the WHO's definitions, so a person with a BMI of 30 or more is generally considered obese. A person with a BMI between 25 and 30 is considered overweight, and a perfect or "normal" BMI is 18.5 to 25 (a BMI less than 18.5 is considered underweight). But due to emerging science, these definitions are still a moving target in the medical community. For years we thought that an optimal BMI for the general population was around 23. But then Dr. Katherine Flegal's explosive paper came out in the beginning of 2013 in *JAMA: The Journal of the American Medical Association*, reporting on nearly three million people who showed that the ideal BMI is actually between 25 and 30. People in this "overweight" and "mildly obese" category had the lowest mortality rate—beating out even those considered "normal" or "healthy" weight.

Unfortunately, BMI is a crude way of determining one's risk for obesity-related illnesses and disorders. And it's far from acting like a crystal ball; it cannot predict who will die prematurely and who will live long and thrive. Each one of us carries excess weight for a variety of reasons, be they genetic, behavioral, psychological, or social, and we may be healthy or unhealthy, die prematurely or live long, for an entirely different set of reasons.

BMI is not as new as we think—it was introduced in the early nineteenth century by a Belgian named Lambert Adolphe Jacques Quetelet, who was a mathematician, not a physician. He devised a quick and easy formula to measure the degree of obesity in the general population for the government's use. A prime example that underscores the problem with BMI, which I've already mentioned: Millions of people complain that as they age, weight loss becomes harder. More to the point, they feel much fatter than they did in their youth, yet they weigh the same. How can this be?

Indeed, the older person *is* fatter than her younger self, even though her BMI hasn't changed. Your height, weight, and BMI may not alter through the years, but your body's *composition* can most definitely reorganize, shifting from muscle to more fat. BMI makes no allowance for the relative proportions of bone, muscle, and fat in the body. Nor does it make concessions for fitness levels and different types of body fat, for the fat that clutches your vital organs—what we call *visceral fat*—is not the same as the fat we attribute to thunder thighs, cellulite, and big butts. Bone is denser than muscle and twice as dense as fat, so a person with strong bones, good muscle tone, and low fat will have a high BMI. Thus, some athletes and fit, health-conscious individuals who work out regularly can find themselves classified as overweight or even obese. Plenty of professional football and basketball players, for example, would be categorized as overweight or obese despite their favorable health status and fitness capacity, and many people labeled as overweight or obese by BMI standards can in fact be healthy by every other measure. (I have personally witnessed many heavy-looking runners breeze past me during races; their outer appearance and BMI number defy their true health as defined by body composition and fitness level.)

Conversely, those with low BMI and little muscle mass may think they win the award for their number (and figure), yet would be surprised by what the research shows, which I just mentioned a few pages ago: Having low muscle mass and low body fat is probably the *worst* body composition to have if you want to enjoy a good, long life. The chief reason is simple: These people tend to have low cardiorespiratory fitness. They are the individuals who try to control their weight (and, ironically, their health) through dietary choices alone. But they miss the proverbial boat big-time. Without the cardiorespiratory fitness, they fail to reap the metabolic benefits that can only be achieved through exercise. And they are likely to carry the kind of invisible body fat that can wreak metabolic havoc on a body.

In addition to having a clear understanding about what BMI denotes, it's important to know the difference between lean muscle mass and thinness. *Lean muscle mass* is the term we use in science to refer simply to your (fat-free) muscle mass. (You may also come across the term "lean *body* mass," which is the more comprehensive category encompassing everything in your body besides fat: organs, bones, blood, muscle, and skin.) For purposes of the discussion going forward in the book, and to avoid confusion of the words *lean* and *thin*, I will refer to just muscle mass when talking about lean muscle tissue. And when we talk about people being lean or thin, we're referring to those who don't carry a lot of extra fat; by BMI standards, they are either underweight or of normal weight. Keep in mind, too, that you don't have to look like a bodybuilder to have high muscle mass. Lots of people with high muscle mass look to be of normal weight, and can even be overweight or obese. It's difficult, however, to have a very low BMI and high muscle mass, since muscle is so much denser than fat.

Although being thin and unfit is worse than being overweight or even mildly obese from a health standpoint, we often idolize, covet, and praise the former while we penalize, berate, and stigmatize the latter. In Abigail Saguy's illuminating book, *What's Wrong with Fat?*, she presents each of the various ways in which fat is understood in our culture today, especially with regard to how we frame fatness in conversations about health. She reveals why we've come to view fatness primarily as something negative, despite considerable scientific uncertainty and debate in many areas revolving around fat. Saguy, an associate professor of sociology and of gender studies at UCLA, makes some very convincing statements, suggesting that public discussions of the "obesity crisis" often do more harm than good, leading to bullying, weight-based discrimination, and misdiagnoses. She writes: ". . . our uncritical reliance on a medical and public health crisis frame of corpulence leads us to emphasize the risks associated with overweight and obesity, while glossing over the health risks associated with 'underweight' or 'normal weight,' as well as those cases where being 'overweight' or 'obese' seems to be protective of health. This begs a social, not a medical explanation." And she's right: The obesity narrative isn't just about the health implications; it's *also* a deeply social conversation. Body size typically intersects with class, race, and gender.

In detailing her argument, Saguy does one thing exceptionally well: pointing out the blatant inconsistencies between the health risks associated with an elevated BMI and people deemed to be in the "normal weight" range. It may be common knowledge that a high BMI is correlated with a greater risk of conditions such as type 2 diabetes and heart disease (and, by extension, the underlying abnormalities, as measured by blood pressure, triglycerides, cholesterol, glucose, insulin resistance, and inflammation). But what's not common knowledge is that almost one-quarter of "normal-weight" people also suffer from metabolic abnormalities. On top of that, more than half of "overweight" and more than one-third of "obese" people look perfectly healthy from a metabolic standpoint when you look at their blood work and lab test results (i.e., they don't have hypertension, insulin resistance, diabetes, or high cholesterol, or show any of the typical red flags we attribute to carrying the burden of extra weight). We call this category, which I think is the ultimate paradox to the layman's vocabulary, *metabolically healthy obesity*. These individuals are indeed obese, but they don't have any disorders that we associate with obesity, and those who have the added bonus of being fit within this category do much better from a health standpoint than normal-weight people with these disorders.

To put this into perspective, in the US there are sixteen million people who are of normal weight yet who have metabolic abnormalities, and there are fifty-six million overweight and obese individuals who have no such abnormalities. One plausible reason for this unexpected discrepancy is that weight isn't the issue making the delineation between these two camps. It's the twin powers of physical fitness and nutrition. We can point to several studies that bear witness to this fact, including some that track physically fit "obese" individuals who have lower incidence of heart disease and death from any cause than do sedentary people of "normal" weight. The impact of good nutrition was most recently showcased in a clinical trial published in *The New England Journal of Medicine*, revealing how adopting a Mediterranean diet reduced cardiovascular risk *independent of weight loss.*

The decision of the American Medical Association to classify obesity as a disease will no doubt put greater pressure on the Food and Drug Administration to approve more weight loss drugs and increase the odds that insurance companies will reimburse their cost, but this move creates a double-edged sword. Labeling obesity as a disease implies that moving into the category of obesity (i.e., under current standards for adults, inching from a body mass index of 29 to 30) is equivalent to contracting a disease. It's one thing to define a disease by a set of symptoms or identifying a specific pathogen. But it's clearly another to define it by a number on a measuring system that's largely arbitrary. What's more, creating such a designation may backfire on the hope of removing the stigma associated with obesity; "the disease categorization may reinforce blame by raising the stakes," says Saguy. She goes on: "If obesity is a disease, parents of fat children may not merely be silently judged as bad parents but also accused of neglect and child endangerment." I agree with her. I also concur that a more productive approach, especially from a public health standpoint, is to focus on the underlying cardiometabolic risk factors that contribute to the real diseases at play here, namely type 2 diabetes and heart disease. Instead of drawing the line between obese and nonobese, we need to acknowledge that in every category of weight there exists both metabolically healthy and metabolically unhealthy individuals. And rather than tirelessly push weight loss per se, we should promote cardiometabolic fitness and encourage people of all sizes to think about their health in terms of how well they eat and exercise rather than a number on the scale (or BMI).

The idea that BMI is a faulty measure of obesity and that obesity is an imperfect predictor of health was further confirmed by a 2009 study that used the National Health and Nutrition Examination Survey to estimate excess deaths for people of standard BMI levels as well as for those with comparable body fat percentage, waist circumference, hip and arm circumferences, waist-to-hip ratio, sum of four skinfold thicknesses, and waist-to-stature ratio. It found *no* systematic differences between BMI and other variables.

Note: Despite the flaws inherent in BMI in terms of health, it's still part of our vocabulary today and will be part of the conversation going forward. BMI has its imperfections, but it's not clear that there are better models for reference points, and it's still used in research circles as a rough guide to individual weight status. We'll be using it mostly to define different weight categories, and what the health implications are for people within those categories based on the new data.

A New Vital Sign

Although many people strive to maintain an ideal weight solely through their eating habits, it is a tough achievement that's often ultimately unsustainable. It takes about thirty-five miles of walking to burn off one pound of fat, whereas a diet program can have a much faster impact via shaving thousands of calories over the course of days or weeks. But in the long run, who can stick with a thousand-calorie-a-day diet? Who wants to starve themselves? And if you're doing this by following a highly restrictive, low-carb protocol that makes you feel deprived, you're less likely to have enough energy to exercise. So your weight ends up fluctuating dramatically, and all the while you lose out on the benefits of exercise. And you probably don't feel so great either. No wonder most diets fail.

I realize that physical fitness is a hard sell. A lot of folks don't enjoy working out, they hate sweating, and they'd much prefer just to control their health with food. But by the end of this book, I hope I will have changed your feelings about exercise. Physical fitness should be a vital sign.

A New Day

I've touched upon a lot of issues in this chapter to set up the balance of the book, which will take you deeper into the story of the obesity paradox and equip you with a new way of thinking about—and looking at—your body. We live in a society that pretty much leads us to believe that the thinner we are, the healthier we are, and that a diagnosis of obesity is the same as a big fat nail in our proverbial coffin. But this perspective is flawed on two counts. First, it's based on ill-defined parameters, in which fat and obesity are misunderstood (and misinterpreted). And second, it leads to the endorsement of lifestyle habits that jeopardize our health and longevity.

If I had to point to one big question raised by the volume of data that we've gathered, it would be this: Are we so obsessed with the obesity epidemic that we're going to extremes and teaching the wrong habits for healthy living? Can we change the course not just of our obesity "epidemic" but of the challenges presented by other chronic diseases simply by redefining fat and fitness in the medical textbooks?

The answer is a resounding *yes*. The obesity paradox may be our generation's window for looking into the body and coming to an exceedingly better understanding of it.

{ CHAPTER 2 }

The Ultimate Scapegoat: The Conventional Wisdom About Obesity as a Modern Scourge

Ask any well-read person on the street what obesity can do to one's health, and you'll hear a litany of challenges that exact a tremendous price: diabetes, coronary artery disease, stroke, and cancer, among others. A doctor would add to this list: insulin resistance, hypertension, high LDL (bad) cholesterol, elevated blood fats (triglycerides), chronic kidney disease, asthma, sleep apnea, osteoarthritis, and depression. Recent evidence indicates that obesity is associated with more diseases than smoking, alcoholism, and poverty. If current trends continue, it may soon overtake cigarette abuse as the leading cause of preventable death in the US.

We saw in the previous chapter how pervasive obesity has become worldwide, affecting millions and costing billions. Many argue that should we fail to stop the obesity epidemic, we will soon lose serious ground on extending our life expectancy—we may actually see our lives getting shorter. But I would temper such an ominous statement by challenging the current definition of *obesity* when it comes to blaming all manner of mortality on it.

The relationship between obesity and myriad illnesses is well documented. Having published hundreds of papers in our most prestigious journals—from the *Mayo Clinic Proceedings* to the *Journal of the American College of Cardiology* and *JAMA*—I've found there's no question that obesity has profound adverse effects on the body that are both direct and indirect.

Before we get to the details of the obesity paradox, I'm going to reveal all that we know from one side of the story: the side that proves obesity is hard on the body. This conversation will include an exploration of two chief areas:

- How a heart in a normal body of healthy weight functions versus how a heart in an overweight or obese body is forced to adapt and function differently.
- How fatness contributes to "obesity-related conditions and illnesses" ranging from arthritis and asthma to cancer.

The Heavy Heart

Obesity stresses the heart in many ways. It leads to high blood pressure, lipid disorders like high cholesterol, and blockages in arteries; it forces the heart to work harder and increase its size to do so. For these reasons, among others, it contributes to the risk of heart failure, coronary heart disease, sudden cardiac death, and atrial fibrillation (irregular heart rhythms). The question is how.

For purposes of this discussion, I'll keep the science as basic and brief as possible. It's important to gain an understanding of the ways in which obesity can influence the structure and function of the heart in order to grasp not only the obesity paradox, but the interplay of the various networks within the human body. What goes on in your blood, fat cells, and metabolism all affects your physiology, the state of your heart, and your risk for illness in general.

Is atherosclerosis (hardening of the arteries)—the disease that causes heart attacks and strokes—a modern plague brought on by smoking, obesity, and sedentary lifestyles? CT scans of 137 mummies spanning four geographies and four thousand years of history show that hardening of the arteries has always been commonplace, especially in older individuals. Such a finding means that a condition like atherosclerosis could just be a part of the aging process rather than a consequence of eating too much steak and potatoes.

Perhaps one of the best ways to understand the complexity of the human heart and its vulnerabilities is to consider the difficulties in our attempts to replicate its machinery artificially as a solution to heart disease, specifically heart failure—our number one killer. Every year, nearly half a million Americans die because of heart failure, and millions of others are severely disabled by it, unable to lie flat in bed or walk to the mailbox without gasping for breath. Right now as many as five million of us suffer from some form of the disease. And there's no quick fix. Transplantation is harder to come by than most people think; only about two thousand usable hearts a year become available and on any given *day* about three thousand people are on the waiting list. As the old saying in medicine goes, "Counting on a heart transplant to cure heart failure is like counting on the lottery to cure poverty." Why the shortage? Healthy, transplantable hearts are increasingly rare finds. They must come from people who were in excellent health but somehow wound up dead (and not for too long). Even when a heart is available, a successful transplant isn't always possible. So it's no surprise that people have tried to fix this inconvenient supply-and-demand reality by building an artificial heart. After all, the heart is just a fancy pump and people have been constructing pumps since the Mesopotamians invented the shadoof to raise river water more than five thousand years ago. How hard could it be?

Very hard. So hard, in fact, that scientists couldn't even fathom replacing the entire heart with a machine until the latter half of the twentieth century.

To get a quick sense of why this is such a demanding feat, try curling a two-pound barbell continuously, without a break. How long can you last? A few minutes? An exhausting hour? An impressive day bound for the Guinness World Records? At first, two pounds might not feel like much, but it can feel like two hundred pounds in a short period of time. Now consider this: Your heart does the equivalent of this exercise twenty-four hours a day, to the tune of thirty-five million beats a year. It expands and contracts ("beats") one hundred thousand times a day, pumping five or six quarts of blood each minute, which is about two thousand gallons daily. And it achieves this without a single pause for as long as you live (or at least until it falters momentarily or permanently during a heart attack). This pulsating device is one of the first signs of life within the womb; it begins beating when a fetus is just three weeks old—when the newly forming life resembles a jumble of cells more than anything else and the brain has yet to make an appearance.

It makes sense that the heart would be among the first things to form; it handles one of the most essential ingredients of life: blood. As the heart beats, it pumps blood through a vast system of vessels we call the *circulatory system*. These elastic, muscular tubes carry blood to every part of the body and total about sixty thousand miles in length in an adult. In addition to delivering fresh oxygen and nutrients to the body's tissues, the circulatory system carries the body's waste products, including carbon dioxide, away from the tissues. This is necessary to support the life and health of your entire body. Blood flows continuously through your body's blood vessels, and your heart is the indomitable piece of machinery that makes it all possible.

People have tried to make a fake heart out of metal and plastic that can beat like a real one, but nothing has lasted more than eighteen months. Mother Nature, it seems, can't be outsmarted so easily in this department. From a mechanical perspective, the heart is pretty simple. It's a four-chambered, hollow organ that orchestrates two pumps made of muscle tissue. One pump takes the dark blood returning from the body and pumps it to the lungs, and the other takes the red blood returning from the lungs and pumps it to the body. The upper two chambers are the *atria*, and the lower two are the *ventricles*. A wall of tissue called the *septum* separates these chambers, and blood is pumped through them via four heart valves that open and close like little doors to let the blood flow in only one direction. You've probably heard that the heart creates a "lub-dub" in its processing of blood. The "lub" is both pumps, left and right, being actively filled by the beating of the left and right atria (the filling chambers); the "dub" is the simultaneous ejection or squeezing of the left and right ventricles (the pumping chambers).

But from a biochemical perspective, the heart is anything but elementary. It's an incredibly sophisticated engine capable of adapting quickly to its environment, including insults, injury, and changes in the body that force it to make certain adjustments to sustain life.

A prime example of this would be what happens when you stand up quickly from a sitting or resting position, which causes an immediate drop in blood pressure in your brain as the blood rushes downward. If your body doesn't adjust quickly to raise your blood pressure again, you will faint. The heart is the first responder.

Luckily, none of us have to think about making such a rapid correction or wait for our DNA to signal genes to turn on or off. The moment you start standing and the pressure begins to change, sensors in the heart's aorta and in the carotid arteries in the neck trigger a response that revs up the fight-or-flight part of the nervous system, sending a reflexive impulse through your vagus nerve that immediately tells the heart to beat faster and stronger. Blood vessels also constrict to squeeze blood into a tighter space and bring the pressure back up so the brain doesn't lose blood. It's a brilliant mechanism that ensures our survival, requiring zero thought and without a lengthy adaptation process in the body. It's more or less a preprogrammed reflex.

The heart has to be able to make such quick and smart modifications, for life couldn't keep going if the heart were vulnerable to sudden stops or easily distracted from performing its job. And nowhere is this adaptive capability truer than when it comes to shouldering the demands of obesity.

A medley of changes occurs in the heart and cardiovascular system as a whole when the body is carrying extra weight. These changes are structural, functional, and biochemical. Multiple forces can affect the workings of the heart, some of which are direct and others that are the result of intricate indirect pathways—many of which science is still trying to decipher. Obviously, when your weight is in a healthy range, your body operates more effectively overall. It circulates blood efficiently, has an easy time managing fluid levels, and you are less likely to develop health conditions such as diabetes, heart disease, and certain cancers. But when your weight is tipping the scales, the body—and your heart, first and foremost—must rise to the occasion.

For starters, the more you weigh, the more blood you have flowing through your body. The heart has to accommodate the heavier load of pumping extra blood, and to do so it undergoes some structural changes—a process called *cardiac remodeling*. The heart muscle itself gets thicker, and the thicker the heart muscle grows, the tougher it is for it to squeeze and relax with each heartbeat. One of the most common outcomes of this remodeling is called *left ventricular hypertrophy*, which is enlargement (hypertrophy) of the muscle tissue that makes up the wall of your heart's main pumping chamber (the left ventricle). As the workload increases, the walls of the chamber grow thicker, losing their elasticity and pumping with less force than that of a healthy heart.

Over the years, sometimes decades, the heart may not be able to keep up with the extra load, which makes one susceptible to heart failure—a serious condition in which your heart simply can't pump enough blood to meet your body's needs. We know now that excess weight is one of the biggest risk factors for premature heart attacks, even beating out smoking. In 2008, one of my friends, cardiologist Peter A. McCullough, led a study published in the *Journal of the American College of Cardiology* finding that obese people had a first heart attack 6.8 years earlier, on average, than leaner people. Severely obese individuals (with a BMI greater than 40) had a first heart attack twelve years sooner than those who were of normal weight. Another study, published in 2011 in the British journal *Heart*, found that obese middle-aged men were 60 percent more likely to suffer a fatal heart attack than a similar group of men who weren't obese, even after researchers accounted for factors like cholesterol and blood pressure.

One reasoning for this direct correlation between obesity and fatal heart attacks is that obesity itself causes fat cells to release inflammatory chemicals that damage the cardiovascular system. So it's not that being fat causes heart attacks per se; it's the downstream cascade of events that obesity triggers, particularly an ongoing inflammatory response, that precipitates a heart attack. And because obese people often have larger hearts to cope with their larger size, the organ is more stressed and vulnerable during heart attacks. However, later we'll see a strong contradiction among many people categorized as obese who don't always bear these risks. And they may, in fact, live longer and better than their normal-weight and underweight counterparts. In the British study that I just referenced, for instance, the researchers noted that in and of itself, being obese was *not* linked to a higher risk of experiencing a nonfatal heart attack or stroke.

Only recently have scientists really begun to unravel the direct relationships between obesity and heart disease. For the most part, the link has been understood via indirect associations between being overweight or obese and health conditions that ultimately put pressure on the heart to adapt in ways that lead to cardiovascular disease. In fact, the many faces of fat and the übercomplexity of obesity—especially its negative side effects—can best be summed up and illustrated by looking at the connections fatness has to "obesity-related illnesses." Let's take a quick tour of just some of the many biological repercussions of obesity, all of which relate somehow to the heart's health.

Obesity and the Kitchen Sink

The law of compound returns governs obesity's relationship with virtually every health condition for which fatness is often blamed. This law is often used in economic circles to describe how wealth grows: A small snowball (money) rolling down a hill will gather weight (interest earned), which increases its speed (money earning interest), which keeps increasing its size (money earning interest and interest earning interest). We can apply the same law to the topic of health conditions linked to fat: Although one health condition can make for a small snowball, such as high blood sugar, that initial condition can quickly trigger others (e.g., weight gain and high blood pressure), which add to the snowball and generate more momentum as the ball gets going down the hill. As more risk factors accumulate, the snowball gets bigger and moves faster. And within no time, the snowball has become a very large, unyielding boulder—obesity.

Notice as you read about these ailments that it's hard to know which came first: the obesity or another related illness. The old chicken-and-egg scenario pervades when it comes to all things obesity related.

Increased Insulin Resistance

Insulin, as you likely already know, is one of the body's most important hormones. It's the main player in our metabolism, helping us usher energy from food into cells for their use. The process by which our cells accept and utilize the vital sugar molecule glucose, the body's preferred source of energy, is uniquely complex. Our cells cannot pick up glucose going by them in the bloodstream. They have to open their doors to glucose with the help of insulin, which acts like a transporter and is produced by the pancreas. Insulin moves glucose from the bloodstream into muscle, fat, and liver cells, where it can then be used as fuel. Normal, healthy cells have no problem responding to insulin, as their cellular receptors for it are abundant. But when cells are relentlessly exposed to high levels of insulin as a result of a persistent presence of glucose—typically caused by consuming too much refined sugars and simple carbs from processed foods—our cells adapt by reducing the number of receptors on their surfaces to respond to insulin. This causes our cells to become desensitized or "resistant" to insulin, ultimately causing insulin resistance. Once the cells are in this state, they ignore insulin and fail to fetch glucose from the blood. As with most biological processes, there's a quick comeback in the hopes of rectifying the problem. And in this case, the pancreas steps up to the plate and pumps out more insulin. So now higher levels of insulin are required for glucose to enter cells.

Of course this creates a recurring cycle that eventually can culminate in type 2 diabetes. If you're a diabetic, by definition you have high blood sugar because your body cannot transport critical glucose into cells, where it can be safely stored for energy. And if it remains in the blood, that sugar will inflict a lot of damage, the heart being among its victims. Diabetes is a leading cause of early death, coronary heart disease, stroke, kidney disease, and blindness. Most people who have type 2 diabetes are overweight; hence the association is made, even though chronic blood sugar imbalances likely started the whole process toward diabetes regardless of weight. Throughout this chain of events, inflammation also runs rampant in the body, a deadly process that further damages the heart, as we'll see shortly.

Insulin does a lot more than just escort glucose into our cells, for it can also be seen as a co-conspirator to the biological events that occur when blood sugar levels cannot be managed well. Insulin is an anabolic hormone, meaning it stimulates growth, promotes fat formation and retention, and encourages further inflammation, elevated blood pressure, and cardiac enlargement. When insulin levels are high, other hormones can be thrown off balance, either by getting revved up or turned down, resulting in adverse consequences. And all of these imbalances in turn push the body further toward a biological cliff, if you will. If it's shoved off that cliff, the body's ability to recover its normal metabolism and support a healthy heart may be forever crippled. Note that such a tragedy can happen regardless of weight, occurring in both skinny and fat people.

WHAT IS "METABOLIC SYNDROME"?

You often hear about this condition in terms of obesity-related illnesses. It's often referenced when obesity is connected to things like high blood lipids (e.g., triglyceride), insulin resistance, and high blood pressure. *Metabolic syndrome* is simply the name for a group of risk factors that raises your risk for heart disease and other health problems, such as diabetes and stroke. In addition to high levels of blood fats and hypertension, additional risk factors include a large waistline, low HDL (good) cholesterol, which helps remove bad LDL cholesterol from your arteries, and high fasting blood sugar, an early sign of diabetes. Although you can have any one of these risk factors by itself, they tend to occur together. People with at least three of the risk factors are diagnosed with metabolic syndrome. In general, a person who has metabolic syndrome is two to ten times as likely to develop heart disease and five times as likely to develop diabetes as someone who doesn't have metabolic syndrome. Mind you, these aren't the only risk factors for heart disease. Smoking, inactivity (which itself is a risk factor for metabolic syndrome), and plain old genetics are also strong risk factors, but they aren't classically categorized under metabolic syndrome. Of all the risk factors I've noted, being overweight or obese dominates the syndrome, since that one condition alone can support and maintain the presence of all the other risk factors. But it's not necessarily a shoo-in for a diagnosis of metabolic syndrome. Numerous overweight and moderately obese people carry no other signs of metabolic syndrome other than their weight. They are metabolically healthy despite the number on the scale. And as we'll see in an upcoming chapter, their secret is often fitness.

High Blood Pressure (Hypertension)

Blood pressure is the force of blood pushing against the arterial walls as your heart pumps blood. If this pressure rises above normal and stays high over time, it can damage the body in many ways, especially the heart and cardiovascular system. Your chances of having high blood pressure are markedly greater if you're overweight or obese. But you can be thin and also have high blood pressure; the condition isn't always about obesity. Because so many people have high blood pressure without knowing it—many of them of normal weight—it's often called the "silent killer."

Coronary Heart Disease

As your body mass index rises, so does your risk for coronary heart disease (CHD), a condition infamous for the waxy, fatlike substances called *plaque* that build up inside the coronary arteries. These arteries supply oxygen-rich blood to your heart, but plaque can narrow or block them, reducing blood flow. Chest pain (angina) or a heart attack can occur as a result. The most influential risk factor for CHD isn't obesity, however; it's age: Over 83 percent of people who die from coronary heart disease are over age sixty-five. Smoking also contributes mightily to coronary heart disease. Not only do smokers endure two to four times the risk of developing CHD than nonsmokers, but they also have a much higher risk of a resulting heart attack. Other risk factors for CHD besides obesity (and often in the absence of obesity) include high cholesterol, type 2 diabetes, high blood pressure, physical inactivity, stress, and certain inherited genes.

Stroke

Plaque buildup in your arteries doesn't just threaten the heart. It can also cause a stroke if an area of plaque ruptures, causing a blood clot to form that can be carried to the brain, where it can block the flow of blood and oxygen. Every forty-five seconds someone has a stroke in the US. That's seven hundred thousand strokes a year, half a million of which are first occurrences, while the rest are repeat strokes. The risk of having a stroke rises as BMI increases, but the top three risk factors are in fact high blood pressure, diabetes, and smoking—factors that apply to many thin or normal-weight people. Higher BMI is also a major cause of atrial fibrillation, a common heart rhythm disorder that dramatically increases the risk of stroke—not from artery blockages but from clots breaking off from the cardiac chmabers impacted by the rhythm disturbance (discussed below). Not everyone who suffers from a stroke is overweight. New research is showing that one strong predictor of risk for stroke is low levels of HDL (good) cholesterol. This explains why young, seemingly healthy people can suffer from a stroke.

Abnormal Blood Fats

If you're overweight or obese, you're much more likely than a normal-weight person to have abnormal levels of blood fats, such as high levels of triglycerides and LDL cholesterol, and low levels of HDL (good) cholesterol. Abnormal levels of these blood fats are risk factors for heart disease. But consider the following: As many as thirty million Americans who have abnormal blood fats also have a normal BMI. They are part of a growing group of "thin-but-fat" people who look slim and are of normal weight but have signs of cardiovascular disease or metabolic syndrome.

Atrial Fibrillation

Obesity is a known risk factor for atrial fibrillation, which is an irregular and often rapid heart rate that commonly causes poor blood flow to the body. During an episode of atrial fibrillation, the heart's two upper chambers (the atria) beat out of sync with the two lower chambers (the ventricles). Symptoms include heart palpitations, shortness of breath, and weakness. This chaotic, irregular rhythm causes many adverse symptoms and dramatically increases the risk of stroke. Aside from obesity, other, often stronger risk factors include advanced age, high blood pressure, excessive alcohol intake, another underlying abnormal heart condition, and, of course, genetics. These risk factors can exist regardless of obesity or even being slightly overweight.

Cancer

These days, it seems like virtually anything questionable in our environment, diet, and lifestyle can be traced to an increased risk for cancer. After all, cancer remains one of our most formidable challenges in medicine, the ultimate bogeyman in health, which begs to be fully understood. In many respects we are no better at treating or curing cancer today than we were several decades ago. And the obesity epidemic certainly adds another layer to the conversation, albeit a controversial one. Today we can associate an increased risk of several types of cancers in different organs with obesity: esophagus, pancreas, colon and rectum, endometrium, kidney, thyroid, gallbladder, and breast cancer after menopause. The exact mechanisms by which obesity can fuel these cancers are still largely unknown and extraordinarily complex (consider this: if we knew the precise mechanism, we'd be better at preventing, treating, and curing cancer). And I should remind you that these are associations, not causal links, for cancer likely has strong genetic roots as well. Obesity probably just adds another ingredient to the mix that tips the scale in favor of cancerous growth. Nonetheless, we have plenty of evidence to show that a host of events likely takes place in the overweight or obese body to support cancer growth. The two most significant factors:

- *Imbalanced hormones due to excessive fat:* High levels of estrogen pumped out by fat cells expose the body to too much estrogen, which in turn feeds cancerous cells, especially those related to sex organs such as the breast, uterus, and ovaries. What's more, obese people often have increased levels of insulin and insulin-like growth factor 1 (IGF-1) in their blood, which may promote the development of certain tumors. Other hormones secreted by fat cells can also skew the body's normal cell growth functionality, stimulating or inhibiting cell growth, which in turn interferes with the body's self-regulating mechanisms to prevent cancer. Cancer is fundamentally a disease of misguided growth. If cells lose their ability to control the rate and speed at which they multiply, cancer looms. And such a shift can be the result of hormonal changes. We know that fat cells can act on substances in the body responsible for tumor growth.

So when there's too much fat around, those substances can gain the upper hand.

- *Altered immune response and inflammation:* Although the immune system is itself a carefully self-controlled network in the body, a minor malfunction can let cancer cells thrive. And a sudden malfunction can be caused by any number of things, including the presence of excessive weight and perhaps some high-risk DNA that has a low tolerance for shouldering such weight. We can't forget the impact of inflammation, either, which is part of the immune system. A body exposed to chronic inflammation, even if that inflammation is low-grade, will be rendered vulnerable to cancerous growths.

The older you are, the higher your risk for cancer of any kind, despite weight. Once we reach middle age our body's ability to combat cancerous growths weakens naturally and inevitably. And as we'll see in chapter 5, overweight and moderately obese people diagnosed with cancer live longer than their thinner peers.

Osteoarthritis

Osteoarthritis is a common joint problem of the knees, hips, and lower back. The condition occurs if the cartilage tissue that protects the joints wears away. The basic pathology behind obesity and osteoarthritis is simply that the weight burden borne by the body and its joints triggers inflammatory pathways that manifest in arthritis. The more weight you carry, the harder it is for your joints to work properly. Hence, even small changes in weight can significantly affect the joint pain experienced by an individual. Let's consider osteoarthritis of the knee. Just ten extra pounds can increase the force on the knee by thirty to sixty pounds with each step (studies have shown that a weight loss of just eleven pounds decreases the risk of developing knee arthritis by 50 percent). Moreover, fat tissues can emit substances and hormones that affect joint cartilage and further spur inflammation that is felt in the joints. Leptin, for example, is a key hormone in weight control, but it's also one of the hormones associated with obesity-induced knee osteoarthritis. It's another case of having too much of a good thing, which turns bad. Genetics are likely at play as well, since there are plenty of thin people walking around with arthritis (and heavy people with healthy joints). Again, this is due to inflammation that's triggered by other mechanisms in the body unrelated to weight.

Sleep Apnea

Sleep apnea is a condition in which a person stops breathing momentarily and frequently during sleep, or the breaths are so shallow that at some point the individual has to gasp for air, causing them to move out of deep sleep. This impacts the overall quality of rest, especially since it reduces the chance of experiencing full sleep cycles. Sleep apnea is said to affect about twelve million American adults, although many are not diagnosed. Obstructive sleep apnea is the most common form of the disorder, in which the airway collapses or becomes blocked during sleep; central sleep apnea is less common and originates in the brain, where the area that controls breathing doesn't send the correct signals to the breathing muscles.

Every time a person with sleep apnea stops breathing during the night, the amount of oxygen in the blood decreases, which means the cells in the body are not getting the oxygen they need. Heart rate increases in an effort to raise blood oxygen levels and deliver oxygen to cells. The repeated resulting biochemical reactions cause stress hormones like cortisol to rise, bathing the body in pro-inflammatory substances that will further stymie normal physiology and create fertile ground for myriad medical conditions (including obesity: cortisol tells fat cells to hold on to their contents). Although many people think of obesity as causative of apnea, that's not always the case.

Some studies suggest that sleep apnea causes hypertension, and it's also strongly correlated with the entire basket of cardio-related conditions such as congestive heart failure, stroke, coronary artery disease, heart attack, and irregular heartbeat. It's also related to headaches, memory problems, mood swings, feelings of depression, and even impotence through several indirect pathways. New research indicates that sleep apnea is behind some people's experience of sudden cardiac death, which kills 450,000 people a year in the United States alone. As its name implies, sudden cardiac death occurs when the heart unexpectedly stops beating due to problems with its electrical system. (This is different from a heart attack, which strikes when there is a blockage in one or more of the arteries to the heart, preventing it from receiving enough oxygen-rich blood.) The electrical glitch is commonly brought on by irregular heartbeats, which can be triggered by sleep apnea, and which in turn can be instigated by obesity. It's another classic case of a vicious cycle.

People with sleep apnea are not always obese, but many of them are, and they also bear other cardiovascular risk factors that could be related to fatness or not. Having these other risk factors already increases a person's chances of sudden cardiac death, so the sleep apnea may just be the tipping point. The compound effect of all risk factors adds up to a bad outcome (death). We actually don't really know whether or not sleep apnea causes heart disease. But we do know that if you have sleep apnea today, you run a much greater risk of developing hypertension in the future. One of the challenges in defining the relationship between sleep apnea and heart disease is that people with sleep apnea often have other diseases as well that muddy the picture and complicate studies. Case in point: If I treat people with high blood pressure or heart failure who also happen to have sleep apnea, their heart conditions improve significantly while their sleep apnea may not. So it's anybody's guess as to the clear association between sleep apnea and heart disease alone.

Asthma

The link between obesity and asthma may not seem so clear, but researchers looking for the reason behind the rise in asthma cases alongside obesity have surmised that the low-grade inflammation that occurs throughout the body with obesity may be a factor. Some have also argued that insulin resistance is another reason for the link. This is debatable, however, and we need to conduct more research. The hard part is separating allergies from asthma, which are related in some people and can be associated with obesity.

In a 2010 study of nearly forty-five hundred men and women from the National Health and Nutrition Examination Survey, about a third of which were overweight and another third were obese, 41 percent had some type of allergy, while 8 percent had asthma. Twelve percent of the obese individuals had asthma, compared to 6 percent of the normal-weight individuals. And the likelihood of asthma rose as body mass index increased and waist circumference expanded. The risk of asthma was more than tripled among the most obese individuals compared to normal-weight people. Thirty-seven percent were either diabetic or insulin resistant. Surprisingly, the study did not find any evidence that insulin resistance was responsible for the relationship. Moreover, the presence of an allergy was not related to either weight or insulin resistance. The researchers pointed out that such findings don't rule out the possibility that insulin is a link between obesity and asthma, or that insulin resistance comes long before obesity sets in. This is a key statement, for it's imperative that we acknowledge that in many of these "obesity-related conditions," obesity is an end result that can further exacerbate a preexisting condition. Put simply, obesity can often be a consequence of—not a reason for—the development of a certain condition.

Kidney Disease

Obesity and kidney disease go hand in hand for one big reason: The kidneys filter the blood to produce urine, so if there's anything going on in the body that negatively affects how the kidneys perform, they can easily become abused. Indirectly, obesity increases the major kidney disease risk factors, such as type 2 diabetes and high blood pressure. Directly, any of these conditions would force the kidneys to work harder, filtering above their normal level (called *hyperfiltration*). This increase in normal function alone is associated with a higher risk of developing chronic kidney disease in the long term. About 10 to 15 percent of the US adult population has chronic kidney disease—one in every nine American adults—although many don't know it yet. While it's common to picture dialysis or kidney transplantation when thinking about kidney disease, in reality it's much more subtle than that on the outset, and the majority of people are unaware they have the condition. And, as with so many other conditions, once chronic kidney disease sets in, it becomes a source of other complications including high blood pressure, low blood count, bone disease, and nerve damage—all of which share associations with obesity directly and indirectly. Indeed, obesity can ultimately *result* from kidney disease, not the other way around. And there are lots of thin people walking around today with all the risk factors for kidney disease with the exception of one: extra fatness.

The previously described conditions are just a sampling of the health challenges aggravated by excessive weight. Although obesity is nowadays blamed for everything but the kitchen sink, in truth it *is* the kitchen sink. It's a center of gravity of sorts from which we can draw an epic assortment of associations, however remote. But because of this, in many ways obesity is also the ultimate scapegoat—the villain we can easily blame when there's anything wrong going on in the body. And we often blame obesity as the prime suspect even when it's a mere consequence of other problems going on in the body. We can find relationships with obesity and its negative effects on virtually every system and organ in the body (which makes its positive effects on people with certain conditions all the more interesting).

Most all of obesity's adverse associations can be traced back to the functioning of the heart. Whether they have a direct or indirect relationship with the cardiovascular system, one fact is clear: The presence of high body fat changes our entire chemistry, down to the ways in which our hearts beat and pump blood through the body. At the center of all these chemical changes is a topic we must address prior to leaping into the details of the obesity paradox itself. That topic is inflammation.

Inflammation's Dark Side

We've known for some time now that the cornerstone of most chronic conditions, from diabetes and obesity to cancer and dementia, is inflammation. All of the conditions I just outlined share direct relationships with inflammation, much more so than they do with obesity. It's been a buzzword over the past decade, even though people's general sense of it is confined to the idea that inflammation relates to the body's natural reaction to an injury. Inflammation explains the tender redness of a cut or insect bite as well as the pain of a sprained ankle or arthritic joint. Although it's often cast under a dark light, inflammation is necessary for survival, serving as an indication that the body is trying to defend itself against something it believes to be potentially harmful. Without inflammation, we wouldn't be able to combat foreign invaders like bad bacteria, viruses, and toxins. But you can have too much of it.

Problems arise when the inflammatory process turns on and stays on, disrupting your immune system and leading to chronic illness and/or disease. It's like having your heat turned on to keep you warm and comfortable, but if it doesn't switch off once a certain temperature is reached, then your environment is going to get uncomfortable and hazardous. Persistently high temperatures will eventually cause things in that environment to become adversely affected. And a lot of people today are living in a sauna—their bodies are besieged by systemic inflammation that's underpinning their chronic conditions. A variety of chemicals are produced when inflammatory processes are continuously turned on. These chemicals can be directly toxic to cells, leading to a reduction of cellular function followed by cellular destruction. Markers for inflammation can easily be detected in the blood, and include such compounds as C-reactive protein. If your levels of C-reactive protein are high, for example, your risk for a medley of health conditions is increased. Since inflammation is believed to have a role in the pathogenesis of cardiovascular events, testing for its markers is an established method to predict who is likely to suffer from heart disease.

It's pretty easy to appreciate how uncontrolled inflammation underlies a problem like a backache from a torn muscle, for example. After all, the first line of drugs used to treat most back pain, such as ibuprofen and aspirin, are marketed as anti-inflammatories. Antihistamines, for another example, are used by allergy sufferers to combat the inflammatory reaction that occurs in the presence of an irritant. I'm happy to see that today more and more people in the lay public are beginning to understand that coronary artery disease, the leading cause of heart attacks, may actually have as much to do with inflammation as it does with high levels of bad cholesterol. Which is why aspirin, in addition to its blood-thinning properties, is often promoted as being incredibly powerful in reducing risk not only for heart attacks but also for strokes and cancer risk.

The strong relationship between inflammation and obesity, and, in particular, fat cells, is subtler than the link we can visualize and likely feel when we have a skin abrasion or achy back. But "fat inflammation" is responsible for a lot of the conditions we attribute to obesity. It's as much a key player in diabetes as it is in hypertension, even though both of these conditions come with separate symptoms and are categorized differently (one relates to metabolism; the other, the cardiovascular system).

No conversation about inflammation would be complete without talking about the other buzzwords of late: *free radicals*. These little molecules act like assaultive bullets in the body against DNA, and they lie at the center of chronic inflammation (and aging in general).

The more technical term for this unavoidable process that affects all living things is *oxidative stress*. In broader, unrefined language, it helps to think of it as a biological type of rusting that can happen anywhere. On your skin, it causes premature aging; on your delicate insides, it can harden blood vessels, damage cell membranes, cripple normal chemical reactions, and essentially wreak havoc on your tissues and organs. As DNA mutations accumulate that cannot be repaired, cellular functions are compromised to the extent there's an increased vulnerability to disease.

As with inflammation, oxidation is a normal part of life. When we turn calories from food into usable energy for the body via our metabolism, oxidation is part of the process; free radical production is inevitable. Tiny, specialized structures or "organelles" in our cells called *mitochondria* are responsible for energy production and release these free radicals while converting nutrients to energy. The body has built-in methods for handling free radicals (and to some degree, free radical production can benefit the body and attack bad cells), but when there's too much that cannot be properly balanced out, ordinary body activities and healthy cells can become impaired and irreversibly damaged. Once tissues and healthy cells become oxidized, they cannot fulfill their duties. People with high levels of oxidation (who typically have high levels of inflammation) often report an extensive list of symptoms ranging from general fatigue and low resistance to infection to muscle weakness, joint pain, mood disorders, and trouble with weight.

Clearly, if you want to live long and avoid the ills of a chronic condition, the goal should be to minimize oxidation and, in turn, inflammation. That's partly why antioxidants are touted as so important; they help counterbalance harmful free radicals. I should also point out, however, that the story of aging isn't all about free radicals (despite what purveyors of antiaging products will tell you). The aging process is also defined by how fast your body can create new cells, and we know that cell division, a key step in generating fresh cells, has its own limitations as we get older. And we don't know for sure the relative contributions of oxidation and limited cell division to mortality. That's still a very active area of study, one that may remain a perplexing field for years or decades to come.

But one thing is certain: Controlling inflammation can go a long way toward helping one stave off illness and disease and preserve metabolic health. In fact, reduced levels of inflammation may explain how some obese people are able to stay metabolically healthy. In general, obesity is linked to a higher risk of diabetes and heart disease, but as I mentioned briefly in chapter 1, some people who are obese never develop high blood pressure and unfavorable cholesterol profiles—factors that increase the risk of metabolic diseases. This phenomenon, described as *metabolically healthy obesity*, makes up as much as 35 percent of the obese population. In 2013, a study done in Ireland and published in the Endocrine Society's *Journal of Clinical Endocrinology & Metabolism* showed that metabolically healthy people—both obese and nonobese—had lower levels of several inflammatory markers. Regardless of their body mass index, people with good inflammatory profiles also tended to have healthy metabolic profiles.

That study was soon followed by another, in the journal *Diabetologia*, that suggested a potential reason behind those who are significantly overweight but escape metabolic problems: better functioning mitochondria. The researchers, from the University of Helsinki, found that obese people who are metabolically unhealthy have impaired mitochondria and a reduced ability to generate new fat cells, and that this might explain why fat cells in unhealthy obese individuals balloon to the point that the cells' internal machinery is severely handicapped and they die off. This is accompanied by inflammation, and it leads to an accumulation of fat where it doesn't belong—in the heart, muscle tissue, and liver, where it's metabolically damaging. Fat cells in the healthy obese, on the other hand, can generate new cells to store excess fat. And they seem to do so just under their skin, where it's fairly harmless. What we need to figure out next is whether the inflammation is the initial troublemaker that handicaps the mitochondria or if the mitochondria malfunction first, leading to inflammation.

The studies don't end there, however. Work is also under way to understand the role of viruses and gut bacteria in contributing to body weight and disease. Our intestines harbor billions of bacterial colonies that collaborate with our digestion and even immune function. Some of these colonies have been implicated in the obesity epidemic while others have protective benefits against weight gain and metabolic disorders. What's more, some of the newest studies are looking at yet another possible mechanism for how obesity causes disease all over the body: exosomes. Small sacs produced by fat, exosomes can be sent off to other parts of the body where they relay disruptive, disease-triggering messages. But this research is still in its infancy, adding another layer of complexity to the story of fat that future research will hopefully figure out. Mind you, exosomes aren't produced just by fat, and they aren't always bad. They are simply cell-derived vesicles that are likely found in all biological cells and play a key function in numerous biological activities, be they ones that contribute to health or dysfunction and disease.

Studies like these further compel us to start looking elsewhere for an understanding of what obesity means and how it plays into our definition of health. We often hear that belly fat is the worst kind to carry, principally because it's an origin for pro-inflammatory compounds and biochemical reactions that end up fanning the flames of inflammation. When it comes to fat real estate, location matters. But, as you're about to find out, it's not everything.

{ CHAPTER 3 }

Fat Real Estate: Does Location Matter?

The road to obesity is anything but straightforward. Interestingly, as scientists further their understanding of obesity and all of its masked faces, including why it can trigger multiple health conditions in some people but spare others, we've uncovered some very intriguing new facts about fat. Fat cells are not all created equal, and different fat deposits have different functional characteristics in terms of how effortlessly they store and give up fat. These different characteristics may help explain why some types of fat feed illness and dysfunction while others are harmless or even prevent disease.

When I was in medical school, the prevailing wisdom was that fat cells were primarily biological storage bins for excess calories. In other words, fat cells were seen as innocuous, inactive cellular warehouses for sidelined energy. But that was a grossly misguided perspective. Today we know that fat cells do much more than simply shelter "silent" calories. Masses of body fat form complex, sophisticated hormonal organs that are very much involved in human physiology. And they are anything but passive, especially when we consider location.

As they say in real estate circles, location matters; it can mean everything. Where your body stores fat can have a major effect on your heart health, as body fat distribution seems to play an important role in the development of obesity-related conditions ranging from heart disease and stroke to some forms of arthritis and cancer. Lugging around too much abdominal fat puts you at greater risk for disease than fat stored on your bottom, hips, upper arms, and thighs. And we can point the finger at our sex hormones for largely controlling how our body fat gets allocated. Estrogens and androgens appear to be responsible for holding the remote control to our body fat throughout much of our lives. And how and when they get produced is dictated mostly by our age and hormonal stage in life.

In women of childbearing age, the ovaries make the highest amount of estrogen compounds, which induce monthly ovulation during an active menstrual cycle. In men and postmenopausal women, estrogens do not come principally from the sex glands. Instead, the body relies on fat cells to produce this hormone, and in postmenopausal women the levels are much lower than those pumped from premenopausal ovaries. Young men produce high levels of androgens, such as testosterone, in their testes, and as they age, those levels decrease. As you can imagine, these natural shifts in sex hormone levels over time in both men and women are associated with changes in body fat distribution and various risk factors for disease. For example, obese older women who, as a result of their extra fat, have more fat-produced estrogen circulating in the body are at an increased risk for breast cancer. But estrogen coming from the ovaries, which is typical in younger, premenopausal women, may not equate with a substantial increase in breast cancer risk. This validates the idea that where estrogen gets produced (and, by extension, where fat is located) is important.

Before menopause, women are more likely to be pear shaped, because they tend to store more fat in their lower bodies. Postmenopausal women and older men, on other hand, tend to have apple-shaped bodies, due to increased fat storages around their abdomens. Taking estrogen supplements after menopause usually helps a woman avoid accumulating belly fat; animal studies tell us that a lack of estrogen usually leads to excessive weight gain. We also know that changes in growth hormone production as we age affect body fat. Growth hormone, pumped out by the pituitary, is essential during our younger years, when we're rapidly developing—growing taller and building bone and muscle. But levels begin to wane the older we get, and this naturally slows down our metabolism. Obese people typically suffer from abnormally low levels of growth hormone, likely due in part to higher levels of hormones (such as cortisol) that antagonize normal production of growth hormone and inhibit how growth hormone functions. (And new studies are showing that growth hormone can in fact prompt weight loss.)

As you'll learn in part II, your body's collective fat mass could very well be one of its most industrious organs, serving a lot of functions beyond keeping you warm, cushioned, and insulated. But some types of body fat do more harm than good. This is especially true of visceral belly fat—the fat surrounding the liver and other abdominal or visceral organs, such as the kidneys, pancreas, heart, and intestines. Visceral fat has also gotten a lot of press lately: We know now that this type of fat is the most devastating to our health. We may complain about our fat thighs, arms, and cellulite, but the kind of fat targeted by researchers today is the kind that's wrapped snugly around our organs deep inside.

What is it about visceral fat that makes it a strong measure of disease risk? Although scientists are still working out all the details in rigorous studies, there are plenty of clues gathering to point us toward some answers. But to understand these clues, you must first understand the different types of body fat and how location dictates their impact on the body.

Essential Fat vs. Storage Fat

From a very broad standpoint, body fat can be divided into two categories: *essential fat* and *storage fat*. Essential fat is just that—it's necessary for normal, healthy functioning and is found in relatively small amounts in your bone marrow, organs, central nervous system, and muscles. In men, essential fat constitutes approximately 3 percent of their body weight, whereas women's is about 12 percent. This is because women's essential fat also includes what's called *sex-specific fat*: Critical for normal reproductive function, it's found in the breasts, pelvis, hips, and thighs. Storage fat, on the other hand, is the fat you accumulate beneath your skin (subcutaneous fat), in your muscles, and in specific areas inside your body. It also includes the fat that protects your internal organs from injury. Men and women generally have similar amounts of storage fat.

Excess visceral, or belly, fat is the classic sign of being overweight and susceptibility to many health risks. Sometimes you'll see this referred to as *abdominal* or *central* obesity. This type of fat actively releases fatty acids, inflammatory compounds, and hormones that ultimately lead to higher bad cholesterol, triglycerides (blood fat), blood glucose, and blood pressure. One of the longest-standing explanations for visceral fat's toxicity has been that it's related to an overactive stress response in the body. The effects of this include raised blood pressure, higher blood sugar levels, and increased cardiac risk. But a newer explanation relies on the concept of lipotoxicity.

Unlike other body fat, visceral fat cells are unique in that they release their metabolic products directly into what's called the *portal circulation*—the passage of blood from the gastrointestinal tract and spleen through the portal vein to the liver. As a result, visceral fat cells that are enlarged and weighed down by excess triglycerides pour free fatty acids straight into the liver. Free fatty acids in general circulation also collect in the pancreas, skeletal muscles, heart, and other organs. None of these locations are designed to store fat, and the free fatty acids pile up, resulting in organ dysfunction, which impairs regulation of insulin, blood sugar, and cholesterol, as well as normal heart function (which is why this type of fat is associated with diabetes and metabolic dysfunction). And all of this activity fuels inflammatory pathways. Visceral fat does more than just generate inflammation down the road through a sequence of biological events; visceral fat tissue itself becomes inflamed.

According to a Mayo Clinic observational study involving nearly thirteen thousand Americans over about fourteen years and led by my dear friend and colleague Dr. Francisco Lopez-Jimenez, people who are of normal weight but have a high waist-to-hip ratio (i.e., belly fat) have an even higher risk of death than people who are considered obese based on BMI alone. (Note: A "high" hip-waist ratio is defined as above 0.90 for men and above 0.85 for women.) The risk of cardiovascular death was 2.75 times higher, and the risk of death from all causes was 2.08 times higher in people of normal weight with central obesity, compared to those with a normal body mass index and normal waist-to-hip ratio. The increased mortality risk accompanying higher ratios of visceral fat is likely due, at least in part, to increased insulin resistance. And we know that insulin resistance in general is one of the primary problems underlying virtually all diseases. Hence, if visceral fat is associated with insulin resistance and other risk factors, it also accelerates the aging process itself.

Unlike visceral fat deep in the belly, the fat that accumulates around women's hips, thighs, and buttocks retains its contents, preventing it from doing any damage to the rest of the body. They are excellent retainers, sucking up extra energy and storing it ardently, thereby protecting the liver. The fact these cells hold tightly on to their fat is why these areas are so stubbornly difficult to target and scale down in size. But there may be a sound reason why nature makes this fat hard to lose: Studies now show the incredible value of lower-body fat. So while we may despise saddlebags and thunder thighs, take heart; they might be good for you.

This growing body of research, much of which has been done on postmenopausal women (since they typically bear more cardiovascular risk factor due to age and hormonal changes), suggests that carrying lower-body fat can actually help stamp out cardiovascular risk factors. Researchers have figured this out by using X-rays to evaluate fat distribution in women's bodies alongside measurements of blood sugar, cholesterol, and triglycerides. They also have determined that, for reasons still unknown, lower-body fat has a positive effect on triglycerides in those with wider waistlines. And in a 2005 study of 3,035 men and women in their seventies, published in the *Archives of Internal Medicine*, University of Pittsburgh researchers found that fatter thighs are linked to a lower risk of metabolic syndrome, a cluster of risk factors that includes high triglycerides and raised blood pressure. Such findings have led some researchers to suggest that liposuction, which is typically performed on stubborn areas like the hips and thighs, might increase one's risk for heart disease. (And in studies done on animals where fat from the upper legs *replaces* visceral fat, benefits can be noted.) Once again, if there's one thing that all these recent discoveries are teaching us, it's this: Body fat has different personalities we never knew existed and that have everything to do with our health and longevity. We'll be exploring more about fat's personalities in part II.

Does Waist Circumference Matter?

So we've just established that belly fat—the flab that snugs your middle and holds your underlying organs hostage—is the worst kind of fat around. But let me play devil's advocate now and challenge the importance of location. If abdominal fat is so terrible, then why is there evidence to show that some people with higher waist circumferences withstand certain health challenges better than their skinnier comrades with the same conditions?

In 1947, the French physician Jean Vague noticed that people with thicker waists were more likely to suffer from heart disease and die prematurely than people who had trimmer waists or carried more weight around their hips and thighs. Studies done decades later finally proved his observations to be right: Abdominal obesity was strongly associated with an increased risk of type 2 diabetes, cardiovascular disease, and death. This was shown even after controlling for body mass index.

Abdominal obesity is often measured by one of two ways: waist circumference or waist size compared to hip size, also known as the waist-to-hip ratio. Several institutions have defined parameters for abdominal obesity around one or both of these measurements, with different thresholds for men and women. Contrary to what you might think, research has also shown that people who are *not* overweight but have a large waist may be at a higher risk for health problems than someone with a trim waist. Perhaps the best study to refer to in this regard is the well-known Nurses' Health Study, one of the largest and longest-run studies to date, which measured abdominal obesity and examined the relationship between waist size and risk of death in middle-aged women. The study involved forty-four thousand volunteers who were healthy at its start. All of them measured their waist size and hip size.

Sixteen years later, the results were in: The women with the highest waist sizes—thirty-five inches or higher—had nearly twice the risk of dying from heart disease, compared to those who had the lowest waist sizes (under twenty-eight inches). Women with thicker waists had a similarly higher risk of death from any cause, compared to those with smaller waists. The risks increased steadily with every added inch around the waist. The study found that even women at a normal weight were at a higher risk if they were carrying more weight around the middle. Normal-weight women with a waist of thirty-five inches or greater, which has been referred to as *normal weight obesity*, or NWO, had three times the risk of death from heart disease in particular, compared to normal-weight women whose waists were smaller than thirty-five inches. The Shanghai Women's Health Study identified a similar relationship between abdominal fatness and mortality in normal-weight women.

For a long time scientists debated which measure of abdominal fat is better for predicting health risks—waist size or waist-to-hip ratio. But we finally have enough evidence from multiple studies to suggest that both methods do an equally good job of predicting health risks. In 2007, for example, a combined analysis of fifteen studies found that waist-to-hip ratio and waist circumference were both associated with cardiovascular risk. Both types of measurements were also found to be comparatively good in predicting such risk. Other researchers have found that waist, waist-to-hip ratio, and BMI are similarly strong predictors of type 2 diabetes.

It's easier, however, to measure and interpret waist circumference than it is to measure both waist and hip. So for most people, it's quicker to focus on waist circumference alone, and that can suffice. Below is a summary of the abdominal measurement guidelines from the most respected institutions worldwide.

Organization	**Measurement Used**	**Definition of Abdominal Obesity**
American Heart Association; National Heart, Lung, and Blood Institute	Waist circumference	Women: > 35 inches; Men: > 40 inches
International Diabetes Federation	Waist circumference	Women: > 31.5 inches; Men: > 35.5 inches (cut points vary for different ethnic groups)
World Health Organization	Waist-to-hip ratio	Women: > 0.85; Men: > 0.9

Now, here's where the research gets really interesting. I noted previously that the Mayo Clinic found a clear association between higher waist circumference and mortality. Their findings were published in 2012. According to the senior author, Dr. Lopez-Jimenez: "We knew from previous research that central obesity is bad, but what is new in this research is that the distribution of the fat is very important even in people with a normal weight. This group has the highest death rate, even higher than those who are considered obese based on body mass index. From a public health perspective, this is a significant finding."

Indeed, it's an important finding, but here's where I can throw a wrench into the works: An equally large, well-conducted study gives us some competing results to consider. Data from both my camp at Ochsner Medical Center on people with coronary heart disease and from UCLA on those with heart failure shows that individuals with a higher waist circumference actually fare *better* (i.e., they live longer than people with smaller waist circumferences who have those same conditions). Confusing?

We're going to explore this inconsistency in detail coming up next, proving once again that the story about fat isn't written in stone just yet. In doing so, I'll point to the data from a paper that I coauthored with colleagues, which was based on nearly ten thousand coronary artery disease patients enrolled in the Aerobics Center Longitudinal Study. Published in the *Mayo Clinic Proceedings*, its conclusions go a long way toward making sense of all this inconsistent data. As it turns out, the debate can't be all about weight, BMI, or waist circumference; it must include the fitness factor. Fitness has a huge say in the relation of fatness to mortality. Without question, there's a crucial divide between being just fat and being fat *and* fit.

{ CHAPTER 4 }

A Big Deal: You Can Be Fat, Fit, and Remarkably Healthy

Picture three different people in your mind: one is unmistakably obese from a visual standpoint and who cannot walk far or fast without feeling bad physically. The second person is outwardly skinny like a runway model with little, if any, muscular definition. The third individual is unquestionably overweight but can breeze through an aerobics class and hold a plank position for minutes in a yoga class. Question: Which one is "healthier"? Who will likely live longer?

If you thought that the skinny girl would win the race here, think again. The latest research from around the world now proves that having a little more padding and athleticism is a powerful body type to possess if you're trying to beat the odds of getting diagnosed with an illness that will shorten your life. So for all of you out there who carry an extra ten to fifty pounds but kick butt in a fitness routine or favorite sport, listen up! And for those who can fit into a size 2 but you get winded climbing stairs or lifting heavy objects, then you should heed the lessons of this chapter too. What you're about to learn about the hidden wonders of fitness on the body's health and longevity might startle you.

The Sitting Disease

If you've paid any attention this past year or so to the media's coverage of health topics, then at some point you might have heard that "sitting is the new smoking." You may have even read articles suggesting that no matter how fit you are, if you sit for most of the day, you are doomed to experience poor health and meet an early death. This means that even if you work out for an ambitious hour or more a day, you could be putting your health at risk if you're incredibly sedentary the rest of the day (i.e., commuting in your car, working at a desk, hunched over a keyboard, watching TV, etc.). Just as smoking is bad for you even if you get lots of exercise, so is sitting too much. Today's world makes it far too easy to stay in the seated position and get a lot accomplished (other than exercise, of course).

It's not that sitting itself is the bad guy. It's the biological effects that prolonged sitting triggers in the body, negatively influencing things like triglycerides, high-density lipoprotein (the good cholesterol), blood sugar, resting blood pressure, and the appetite hormone leptin (which tells you when to stop eating).

The science showing this link between sitting time and total mortality started to emerge a few years ago when researchers at the American Cancer Society released an eye-opening study, published in the *American Journal of Epidemiology*, that pretty much said sitting down for extended periods poses a health risk as "insidious" as lighting up cigarettes or overexposing oneself to the sun. The people in the study were followed from 1993 to 2006, during which time the researchers examined the participants' amount of time spent sitting and physical activity in relation to their mortality. A second study done at the Baker IDI Heart and Diabetes Institute in Melbourne, and also published in the *American Journal of Epidemiology*, concluded that even two hours of exercise a day would not compensate for "spending 22 hours sitting on your rear end." And yet another study, done by close colleagues at Pennington Biomedical Research Center, spanning twelve years on more than seventeen thousand Canadians, found that the more time people spent sitting—regardless of age, body weight, or how much they exercised—the sooner they died.

According to the research, women in particular seem to be more vulnerable to prolonged sitting's adverse effects. In the study performed by the American Cancer Society, women who sat for more than six hours per day (apart from work time) were 37 percent more likely to die during the time period as those who sat fewer than three hours daily. Men who took to their rears for more than six hours a day (also outside of work) were 18 percent more likely to die from heart disease and had a 7.8 percent increased chance of dying from diabetes compared with people who sat for just three hours or less a day. The association remained virtually unchanged after adjusting for physical activity level. My guess is the disparity between the sexes is partially due to men naturally having greater muscle mass, which confers multiple health benefits and helps deflect the negative impact of sedentariness.

What all this research is clearly showing is that the body shuts down at the metabolic level when stationary. And the logic seems very, well, logical: When you're immobile, your circulation slows down and your body uses less of your blood sugar. This effect, coupled with burning less fat, increases your risk of heart disease and diabetes. Another piece of data coming into view is the impact that being immobile has on certain genes. A key gene called *lipid phosphate phosphatase 1*, or LPP1, could be partly to blame. LPP1 helps to keep your cardiovascular system healthy by preventing dangerous blood clotting and inflammation. But it's significantly censored when you sit for a few hours, meaning it's suppressed from expressing itself. And here's the shocker: This gene is not impacted by exercise if the muscles have been inactive most of the day! Put another way, LPP1 is sensitive to sitting but resistant to exercise.

Just how sedentary are we? In a 2012 study published in the *International Journal of Behavioral Nutrition and Physical Activity*, researchers led by Marc Hamilton, PhD, again at Pennington Biomedical Research Center, reported that—whether or not they exercised the recommended 150 minutes a week—people spent an average of 64 hours a week sitting, 28 hours standing, and 11 hours engaged in non-exercise walking (what we'd call "milling about"). That translates to more than nine hours a day of sitting, no matter how active they otherwise were. Although the research focused on women, it likely reflects what's happening in both sexes. What surprised the researchers was the fact that those who exercised rigorously didn't spend less time sitting. In fact, regular exercisers were likely to make less of an effort to move outside their designated workout time. And we have research now to show that people are about 30 percent less active overall on days when they exercise versus days they don't hit the pavement for a run or go to the gym. Why? Presumably, they think they've had their fill—they've worked out enough for one day.

So beyond separating the exercisers from the non-exercisers, a significant concern should be the general lack of walking, standing, and moving our bodies on a regular basis to counteract all the harm that can result from sitting for the majority of the day. We know that moving routinely throughout the day, even if it's just walking around while talking on the phone, taking the stairs instead of the elevator, or simply making a point to get up every hour for a five-minute stroll or jog in place will have positive biological effects to counteract the poison of excessive sitting.

Although a lot of the talk about the risks associated with too much sitting often revolve around diabetes and heart disease, these aren't the only health hazards. The American Institute for Cancer Research now links extended sitting periods with increased risk of both breast and colon cancers. And adding insult to injury, a 2013 survey of nearly thirty thousand women in the United States found that those who sat nine or more hours a day were more likely to be depressed than those who sat fewer than six hours a day. Again, the biological reasoning makes sense: When you're sitting down, your circulation is reduced, and so is the flow of feel-good hormones to your brain.

I've brought up this issue of "sitting disease" to provide the entryway into a discussion about the power of exercise. If sitting can be so detrimental to health, then that fact alone underscores the value—and necessity—of exercise, especially on a very routine, consistent basis. Clearly, we are not designed to sit. We are made to move . . . and move frequently.

The Science of Exercise

We've known that exercise is good for us for centuries, even if the exact reasons and underlying explanations were unclear. But only in the past decade have we really been able to quantify and qualify the extraordinary relationship between physical fitness and total health through laboratory and clinical tests. This has been possible not only by novel collaborations among various fields of science and medicine, but also the development of many advanced technologies that have facilitated our ability to analyze and understand what happens when we take a walk, go for a bicycle ride, pick up heavy weights, or train for a half marathon.

Since the dawn of humankind, we have been active creatures in a continual pursuit of survival. Only in relatively recent times have we gravitated toward sitting all day. Modern technology has afforded us this privilege because virtually anything we need is available without having to exert much effort, much less get off our butts. Our DNA, however, didn't evolve over the past millions of years to thrive in such an environment. Much to the contrary, it has evolved to be physically challenged to some degree. In fact, one can make the case that our genetic makeup *expects* and *requires* frequent exercise just to sustain life. But as we are well aware, only a small percentage of us abide by that requirement today, and our soaring rates of chronic illness prove it.

The idea that physical prowess has shaped human evolution has kept a lot of anthropologists and biologists busy. Evolutionary biologists Daniel E. Lieberman of Harvard and Dennis M. Bramble of the University of Utah have studied and written extensively on the notion of "survival of the fittest" from a literal perspective. In 2004 the journal *Nature* published one of its seminal articles titled "Endurance Running and the Evolution of *Homo*," in which they posit that we survived through the millennia by virtue of our athletic aptitude. As we evolved from earlier primates that resembled chimpanzees, natural selection drove us to become supremely agile beings. Over time, we developed longer legs, shorter toes, less hair, complicated inner-ear mechanisms to maintain balance and stability, and large brains relative to our body size. Our cavemen ancestors were able to outsmart predators and hunt down valuable prey for food that produced energy for successful reproduction, ultimately ensuring our survival. We've inherited those sporty genes. It's a fascinating concept: We are designed to be active creatures so that we can live long enough to procreate and parent children.

The following rewards of exercise have long been proven scientifically. Proof that working the body has benefits that reach even further than the physical:

stamina, strength, and flexibility

optimal coordination

increased oxygen supply to cells and tissues

restful, sound sleep

balanced hormones

better self-esteem and sense of well-being

muscle tone, endurance, and bone health

release of endorphins, which act as natural mood lifters and pain relievers

stress reduction

fewer food cravings

lower blood sugar levels and lowered risk for diabetes

ideal weight distribution and maintenance

brain health—sharper memory, lower risk for dementia, lower incidence of depression

heart health—lower risk for heart disease, increased blood circulation

decreased inflammation and risk for age-related disease, including cancer

higher levels of energy and increased productivity

. . . among many other potential benefits!

Exercise physiology barely existed when I was in medical school. Today entire new fields of medicine have emerged in this burgeoning area, one of which is metabolomics, a form of health profiling that aims to find metabolic patterns in people that either forecast disease or lower their risk for certain illnesses. We can now analyze blood samples to obtain a chemical snapshot of the effects of exercise. And out of this area of study has come the finding that the fitter you are, the more your metabolism benefits, thanks to dramatic changes that occur during physical movement. Although this may seem obvious based on anecdotal evidence alone, the documented science to back up these claims has proven to be downright astonishing. Here's one case in point: In a 2010 study performed by a team from Massachusetts General Hospital and the Broad Institute of Harvard and MIT, fit people were found to have greater increases in a metabolite called *niacinamide* than unfit people. Niacinamide is a vitamin derivative known to enhance insulin release, which means it's involved with blood sugar control. In fact, this team found more than twenty metabolites that change during exercise, reflecting the effects that being active has on how we ultimately process sugars, fats, amino acids, and even ATP (adenosine triphosphate), the primary source of cellular energy. Some revved up during exercise, such as those involved in processing fat. Those involved with cellular stress decreased. (Another experiment, which studied samples taken from different vascular locations, showed that most metabolite changes were generated in the exercising muscles, although some happened throughout the body. In both experiments, several metabolite changes remained an hour after exercise had ended.) Until recently, some of these metabolites weren't even known to be affected by exercise at all.

Documenting how exercise can trigger changes at the genetic level is one of the most thrilling areas of study taking place today. The idea that exercise can change you down to the expression of your genes is no longer a hypothesis. DNA may be static, but the expression of that preprogrammed coding is anything but fixed. Genes can turn on or off, depending on what biochemical signals they receive from elsewhere in the body. Genes in the On mode express various proteins that, in turn, prompt a range of physiological actions in the body. And genes in the Off mode prevent certain things from happening, which can have a positive or negative outcome.

What the Mass General and Broad Institute researchers found confirmed their hypothesis: Exercise does indeed positively impact genetic expressions related to how we use calories, manage blood sugar, metabolize fat, and overall control our metabolism. Departments of neurology at numerous top institutions have similarly recorded stunning genetic changes in the past several years when it comes to exercise's impact on the brain. Some experts in this field would go as far as to say that physical exercise is one of the most powerful ways to change your genes. Aerobic exercise in particular (also known as "cardio"), which increases your body's need for oxygen and includes activities like running, walking, swimming, and biking, not only turns on genes linked to longevity, but it also targets an important gene that encodes brain-derived neurotrophic factor (BDNF), a protein famously called the brain's "growth hormone." More specifically, aerobic exercise increases production of BDNF, and has been shown to reverse memory decline in elderly humans and actually trigger the birth of new cells in the brain's memory center. Although the neurological benefits of exercise may not seem related to one's weight and metabolic health, it further proves the positive impact that exercise can have on our bodies from a molecular, genetic standpoint. If it's good for the brain, it's good for the body.

The relationship between brain health and metabolic health is much closer than you think. New science is pointing to the idea that brain disease, including Alzheimer's, is a third type of diabetes. It turns out that insulin resistance likely contributes to the formation of those infamous plaques that are present in diseased brains. These plaques are the buildup of an abnormal protein that essentially hijacks the brain and takes the place of normal brain cells. It's all the more telling that those with diabetes are at least twice as likely to develop Alzheimer's disease, which is estimated to affect one hundred million people by 2050, a crippling number for our health care system and one that will dwarf our obesity epidemic.

The positive effects of exercise on our genes are not confined to cardio work, however. In 2008 a team of Canadian and American researchers demonstrated the power of strength training when they evaluated the effects of a six-month program in elderly volunteers age sixty-five and older and found that this type of exercise also can help reverse the aging process at the cellular level. By the end of the trial, the expression of about two hundred genes known to become more or less active with age had changed. The genetic profiles of the elderly volunteers who'd gone through the strength-training exercises resembled those of a much younger group (average age twenty-two). And when the scientists dug a little deeper, they noticed that the genes that had changed were those involved in the functioning of mitochondria. These tiny organelles found inside every cell are the body's workhorses when it comes to energy metabolism. Equipped with their own DNA, they are responsible for converting the energy in food into ATP, which, again, is the molecule that provides chemical energy for physiological processes.

The importance of your mitochondria cannot be overstated. Their age and efficiency have a direct correlation with your metabolic health, and as noted earlier, cutting-edge science is allowing us to begin to see that the difference between obese people who suffer from metabolic problems and obese people who don't could be based on how their mitochondria are functioning. Your mitochondria are key to your physiology and are often direct targets for cellular damage. While what you eat and what you expose yourself to in your environment can affect the extent to which your mitochondria function and how much damage they endure, now you can see that the amount of exercise you get does as well. Moderate-intensity aerobic exercise for just fifteen to twenty minutes, three to four times a week has been shown to increase the number of mitochondria in your muscle cells by 40 to 50 percent. That's not very much exercise for a huge increase in the amount of these little engines that thrive behind the scenes of your energy metabolism.

The research that shows how exercise can change the way genes operate has been a game changer. It has highlighted the processes behind the discovery that exercise promotes health and reduces most people's risks of developing many diseases despite weight. One of the ways we think genes can be changed by exercise is through a process called *methylation*, in which certain chemical compounds called *methyl groups* become attached to the outside of a gene and change how that gene operates. The gene's fundamental structure doesn't change, but how it receives and responds to messages from the body does. What the science is also showing is that this methylation process seems to be heavily impacted by our lifestyle choices, particularly with regard to our eating and exercise habits. Several studies of late have documented how dietary factors can affect the methylation of genes, convincing many scientists to think that differing genetic methylation patterns resulting from differing diets could partly explain why one person develops diabetes and other metabolic diseases while another person doesn't. Scientists have also concluded that exercise has an equally profound effect on DNA methylation within human muscle cells, even after a single workout. What's more, the studies are showing that DNA methylation changes are probably "one of the earliest adaptations to exercise" and drive the bodily changes that follow. In other words, exercise acts like a switch in the body to turn on many channels of health and wellness.

Clearly, multiple effects happen simultaneously when the body is engaged in physical activity—chemically, metabolically, physically, molecularly, and genetically. And many of these reactions become intertwined, giving researchers like me plenty of homework as we try to decipher the sequence of steps in all of these intricate processes. Although we know, for instance, that exercise is a potent anti-inflammatory, it's possible that this effect has both genetic roots (i.e., physical activity stimulates genes that suppress inflammation) and purely physiological ones (i.e., physical activity changes the body's chemistry such that it triggers anti-inflammatory pathways without direct instructions from our DNA). We can measure a lot of the changes from a broad standpoint in the laboratory without having to go as far as examining the genes. Scientists have documented time and time again that C-reactive protein—a commonly used laboratory marker of inflammation—is lower among people who maintain an exercise routine, and my colleagues and I have published many important medical papers on this topic.

There are as many mechanical and biochemical aspects to study as there are molecular and genetic. From a mechanical standpoint, we know that exercise increases our lung capacity so we can take in more oxygen, as well as boosts circulation to deliver nutrients to cells. These two reactions alone have numerous and far-reaching effects that at some point, deep down at the cellular level, can change the codes to our DNA's expression.

If there's one thing we are already figuring out rather quickly, and that is currently changing how we look at people in terms of "health," it's that looks can be deceiving. If we can roll back our biological clocks and positively empower our metabolisms (not to mention virtually every other system in the body) just through the magic of movement, then this is the lens through which we can come to understand the so-called "lean paradox," which refers to the way thin people can look healthy but in fact be suffering from many health problems. Some individuals may try to manage their weight and health through diet alone, shunning fitness, and they experience metabolic consequences similar to if they were morbidly obese.

The Magic of Fitness and Muscle Mass

So with that big-picture understanding, let's look at two different scenarios that help further define true fitness:

- **When thinner means sicker:** The perils of being skinny and unfit (i.e., low body fat and low muscle mass) and the lean paradox: a phenomenon characterized by people who appear slim and trim yet suffer from metabolic problems like type 2 diabetes.
- **When thicker means hardier:** The counterintuitive blessings of *higher* body fat and high muscle mass.

While it's reasonable to see how a rail-thin person with little muscle mass could be unhealthy, the benefits of muscle *and* fat combined are not so obvious. Two facts can help us out here. For one, as body fat increases, muscle mass and strength also tend to increase, generating more health-enhancing benefits. In 2011 the *Mayo Clinic*

Proceedings published a study of mine involving 581 patients after they'd had a major cardiovascular event. The people who had higher body fat (more than 25 percent in men and more than 35 percent in women) and higher muscle mass only experienced a 2 percent three-year mortality (translation: only 2 percent of that group died within three years). But the group with both low muscle mass and low body fat had a 15 percent three-year mortality—that's seven times the mortality of the group with high body fat and high muscle mass. Those in between, the people with high body fat and low muscle mass, or, conversely, those with low body fat and high muscle mass, showed an intermediate mortality. It's intuitive to think that having high muscle mass would be protective and be associated with a better prognosis, but high body fat? And what explains the fact those individuals with high body fat and high muscle mass were more likely to outlive the people with low body fat and high muscle mass?

One strong possibility is that extra calories in the form of fat give someone an advantage when it comes to combating an illness. It's also possible that fat acts in certain ways to counter the adverse effects of a health condition, for there are certain roles that body fat plays in our health and that no other body tissues can replace. However, very likely, genetic forces are involved. While an overweight individual might develop heart disease, for instance, because of the risks created by having more fat, as we have discussed previously, it's conceivable that a thinner person would be likelier to develop heart disease from a genetic propensity—the thin person's prognosis would be worse in comparison to the fatter person's due to the power of those genetics. In other words, an overweight or obese person may not have developed the heart disease in the first place had weight gain been prevented during his adult life, whereas the person who develops the same heart disease—despite maintaining a lean profile, and despite on paper appearing less sick than the heavier person (e.g., lower blood sugars, triglycerides, blood pressure, and inflammatory markers, and higher levels of HDL or "good" cholesterol)—may be more doomed by his particular genetic profile.

We know by now that if you're going to prioritize fitness or weight as you age, it's much better to strive for fitness and be on the thicker side than to be thin and unfit. Loss of fitness is a much stronger predictor of mortality than weight gain. Case in point: Slender, young people can develop insulin resistance and diabetes, a disease traditionally blamed on being overweight in older adults (surprise: about 15 percent of people with type 2 diabetes aren't overweight, according to the National Institutes of Health). Some experts call this condition TOFI—thin outside, fat inside. A person's slender outer appearance will hide the nasty biology taking place deep inside, where the fat is sticking to the abdominal organs instead. As I outlined earlier, this visceral fat in turn causes inflammatory substances to antagonize the liver and pancreas, putting you at risk for type 2 diabetes as your cells lose sensitivity to insulin. So while you might look good, your insides are acting as if you are obese.

This growing trend among thin populations clues us in to just how important it is to exercise *regardless of weight*. Neglecting to exercise and controlling your weight through dietary choices alone can be downright damaging. If you want to lower your blood sugar, breaking a sweat will beat any popular diet. Your muscles will devour glucose at twenty times their normal rate with just moderate exercise; it's been proven that a daily thirty-minute walk at a brisk pace can cut one's odds of developing type 2 diabetes by 58 percent. I don't know any diet that can market such a claim.

Decades-old data from major epidemiological studies has demonstrated the positive impact of cardiorespiratory fitness, which is classically described in the medical literature using what's called estimated *metabolic equivalents*, or METs, typically determined by certain tests on a treadmill. My good friend and colleague Dr. Steven Blair, of the University of South Carolina, is widely considered the "father of aerobics" and is the main author of the now-famous Aerobics Center Longitudinal Study (ACLS), which began in 1970 as an observational study to investigate health outcomes associated with physical activity and cardiorespiratory fitness and has since amassed a database that contains more than 250,000 records from almost 100,000 individuals. He has contributed significantly to the literature: With Dr. Duck-chul Lee and colleagues, Dr. Blair published a 2011 study in *Circulation* using ACLS data showing that an increase of 1 MET in cardiorespiratory fitness over a six-year period is associated with a 15 percent reduction in all-cause mortality and a 19 percent reduction in cardiovascular death risk ("all-cause" mortality is the catchall phrase for "risk of dying from anything"). Although the study examined only men, my guess is similar patterns would be found in women, too. Interestingly, a change in body mass index over that same time period did not impact the risk of death in their analysis once changes in fitness were considered. The men who lost fitness had higher all-cause and cardiovascular death risks regardless of BMI change.

In another study under the direction of Steven Blair that utilized ACLS data, which was published in 2013 in the *European Heart Journal*, researchers showed that people can be obese but metabolically healthy and fit, with no greater risk of developing or dying from cardiovascular disease or cancer than normal-weight people. This particular study was the largest one to date, encompassing more than forty-three thousand people who were recruited to the ACLS between 1979 and 2003. The individuals completed a detailed questionnaire about their medical and lifestyle history, and underwent physical examinations that included a treadmill test to assess fitness levels. They were followed until they died or until the end of 2003.

The results were remarkable: Forty-six percent of the obese participants were metabolically healthy—they didn't suffer from conditions such as insulin resistance, diabetes, and high cholesterol or blood pressure despite their high body fat. And after researchers adjusted for several confounding factors, including fitness, the metabolically healthy but obese people had a 38 percent lower risk of dying from anything than their metabolically unhealthy obese peers. Moreover, no significant difference was seen between the metabolically healthy but obese and the metabolically healthy normal-weight participants. Among the metabolically healthy but obese people, the risk of developing or dying from cardiovascular disease or cancer was reduced by 30 to 50 percent. And best of all, the researchers didn't find any significant differences between these folks and the metabolically healthy normal-weight participants.

The study highlighted two important facts noted by one of its lead authors, Dr. Francisco Ortega. First, that a better cardiorespiratory fitness level should be considered a component of metabolically healthy obese people. And second, that metabolically healthy but obese individuals have a similar prognosis to that of metabolically healthy normal-weight individuals, and a better prognosis than their obese peers with an abnormal metabolic profile. Dr. Ortega summed up the takeaway succinctly when he said to *ScienceDaily*: "Based on the data that our group and others have collected over years, we believe that getting more exercise broadly and positively influences major body systems and organs and consequently contributes to make someone metabolically healthier, including obese people." Indeed, studies like this call attention to the important role of physical fitness as a health marker. What's more, they bring out the need for more clinicians to take into account patients' fitness levels when estimating risk for certain illnesses like cardiovascular disease and cancer, especially in obese individuals.

That same year, I coauthored a paper in the *Mayo Clinic Proceedings* with Drs. Paul McAuley and Steven Blair and colleagues that accentuated the ultimate conclusion from these research findings: Cardiorespiratory fitness greatly changes the relation of fatness to mortality. In our study, we analyzed ACLS data from 9,563 men whose mean age was 47.4 years and who had documented or suspected coronary heart disease. We found that the thinnest individuals had higher mortality risk than did the overweight and the mildly obese. Among those classified as "fit" (meaning they were not in the lowest one-third of fitness for their age and gender), their mortality and risk for cardiovascular death were exceptionally low, and the lean and fit had just as good a prognosis as the heavier and fit. Such surprising results have been confirmed by other research groups as well worldwide.

Although, ideally, reducing your risk for cardiovascular disease would include maintaining both a healthy weight and high cardiorespiratory fitness, substantial data from Drs. Blair, Timothy Church, and others indicates that cardiorespiratory fitness is so protective that it essentially cancels out the adverse impact of traditional risk factors on mortality, including overweight, obesity, metabolic syndrome, type 2 diabetes, and hypertension. So although both how fat and fit you are have a hand in your health and longevity, the current evidence shows that your fitness level may be more important than your body fat level in terms of preventing and managing disease.

The results from research groups (in addition to my own) that debunk old notions about fatness being bad no matter what have been alarming, to say the least. And even though many of these findings are from people who've already been diagnosed with some form of heart disease, they nonetheless have implications for everyone, including the very healthy. More than anything, I think they put the spotlight on the power of fitness and, in particular, muscle mass. If you can increase your muscle mass and cardiorespiratory fitness (which usually go together), then your risk factors for metabolic disorders and diseases of certain organs and cancer can be greatly reduced. Irrespective of fat levels, overall fitness independently enhances overall health; it seems to be the secret weapon against the higher risks born from obesity. This is when we can really say that health is about "strength in numbers."

That said, all too often people focus on cardio fitness, forgetting that greater muscle mass seems to counteract the poor cardiovascular profile of people who are overweight or obese. Just what is muscular fitness (as opposed to cardiorespiratory fitness)? It's pretty straightforward. Muscular fitness entails three main aspects: muscular strength, or the ability of a specific muscle or muscle group to generate force or torque; muscular endurance, the ability to resist repeated contractions over time or to maintain a contraction for a prolonged period of time; and explosive strength, also called "muscular power," which is the ability to carry out a maximal, dynamic contraction of a single muscle or muscle group in a short period of time. So you don't have to look like a bodybuilder or Olympic swimmer to have muscular fitness. Anyone who has kept a strength training routine (whether it's with classic weights, gym machines, or resistance against the body's own weight, as with yoga or Pilates) likely has some decent muscle fitness.

People also fail to appreciate just how valuable muscle mass is to quality of life, longevity, and the ability to optimize metabolic health. Unlike fat, muscle is a highly active tissue that requires the body to pump a lot of energy into it so that it stays in good working order. We have plenty of studies to refer to muscle's positive influence on a healthy metabolism. Researchers at UCLA just recently investigated whether or not increasing someone's muscle mass to average or above-average levels leads to improved blood glucose management (i.e., better insulin sensitivity and lower risk of prediabetes and diabetes). Lo and behold, it does. While low muscle mass has been known for some years to be connected with insulin resistance, now we can say that increasing muscle mass helps prevent type 2 diabetes. This study, which encompassed more than 13,600 adults over the age of twenty, reflected a departure from the usual focus we place on weight loss to improve health. And it clearly spelled out the compelling argument for maintaining fitness and building muscle. In particular, the study's results pretty much established that every 10 percent increase in the ratio of skeletal muscle mass to total body weight is associated with an 11 percent reduction in risk of insulin resistance and a 12 percent drop in the risk of developing prediabetes or diabetes.

The underlying biology for this finding makes sense: As I've noted, muscles are big consumers of blood glucose. Every time we are active physically, our muscles feast on blood sugar, resulting in less glucose circulating in our blood vessels. Meanwhile, having more muscle mass in general appears to protect us against insulin resistance, the cornerstone of diabetes.

Mind you, *muscle* doesn't just refer to your biceps and triceps. Think about the involuntary muscles that move all the time: Your heart is a muscle that pumps oxygen and nutrients to cells; muscle action moves lymph through your lymphatic system, which is an integral part of your immune system. Even breathing and sweating depend on muscles. Muscle is in constant use by the body to keep it alive and well. This is why the more muscle you have, the faster your metabolism will be. It's the main determinant of whether your metabolism is zooming at 100 miles per hour or puttering at a slow 10 mph. Strength training can provide up to a 15 percent increase in metabolic rate. While aerobic exercise burns fat chiefly *during* exercise, strength training utilizes fat hours *after* exercise. The burn keeps going for longer than you're actually working out. (As an aside, we also know that strength training is an effective antidepressant, and can even improve sleep quality.)

Fitness gurus will tell you that strength training becomes more vital the older one gets. And they are right, for it supports muscle mass like no other form of exercise and can help increase not only strength but also bone mass. The muscles you engage when you lift a weight put pressure on your bones, forcing them to get stronger. In fact, recent studies have shown that loss of bone density may be a better predictor of death from atherosclerosis than cholesterol levels.

In most people, muscle strength peaks in our twenties and then gradually decreases. Every year after the age of twenty-five, the average American gains about one pound of body weight, yet loses one-third to one-half pound of muscle. Without strength training, most people experience a 30 percent loss in overall muscle by age seventy. With this loss of muscle strength, we naturally become less active because daily activities become more difficult and exhausting to perform. And women are at an unfortunate disadvantage. By design, men have much more testosterone than women, making it that much easier for a man to build muscle, since this hormone promotes the formation of muscle mass. It's their main sex hormone, produced in the testicles, whereas women's chief sex hormone is estrogen. Plus, recent research suggests that women on average will lose muscle mass twice as fast as men the same age, which can make a huge difference in their ability to maintain an ideal weight. We'll see how loss of estrogen after menopause can further create challenges to one's metabolism, essentially handicapping the body's innate technology for managing appetite and fat storage.

What else is going on when we force our muscles to exert themselves and grow bigger and stronger? To answer these questions, look no further than the most underappreciated and unrecognized aspects of muscle. (Yes, it's nice to look toned and be able to lift heavy objects, but, as you now can appreciate, it's much more than that!) For one, muscle plays a key role in the entire body's protein metabolism—just as fat stores precious extra calories for energy reserves, muscle serves as an emergency supply of the amino acids we need to build tissues, biomolecules, and hormones. To be clear: The body doesn't store amino acids like it does fat and carbs; it will draw from its own tissue by breaking down its protein sources, usually muscle, to meet its need for amino acids if there's not enough coming in from the diet.

In 2006, Robert Wolfe, of the University of Texas, wrote a paper for *The American Journal of Clinical Nutrition* titled "The Underappreciated Role of Muscle in Health and Disease," in which he outlined muscle's role in the body as an unsung hero, drawing attention to muscle metabolism's key role in the prevention of many common pathologic conditions and chronic diseases. His paper dovetails with what my colleagues and I have found when studying the biological benefits of muscle and muscular strength.

More muscular strength seems to be associated with the following:

Less total and abdominal fat; smaller waist circumference

Less weight and fat gain

Lower risk of developing hypertension

Less insulin resistance

Lower chronic inflammation

Less incidence of metabolic syndrome

Lower risk of high blood pressure

Lower levels of triglycerides (blood fats)

Lower levels of bad (LDL) cholesterol

Better blood sugar balance

Virtually all tissues and organs, especially the skin, brain, heart, and liver, require ongoing upkeep and maintenance of their protein content. These essential tissues and organs rely on a steady supply of amino acids via the blood to serve as precursors for the synthesis of new proteins. Since the early 1960s, we've known that when the body is deprived of incoming dietary nutrients, it will turn to muscle protein itself to replace the amino acids in the blood that are quickly used by other tissues. Another way of stating this is to say your body will "eat" your muscle if you're not getting enough protein-building ingredients in your diet. The body needs a constant supply of amino acids to manufacture proteins and to support the liver's glucose-creating engines, which ensure you're never lacking the energy the body needs to run efficiently. By depending on muscle mass in cases of emergency, the bloodstream will never run out of amino acids—provided adequate muscle mass is available.

Studies have shown, for example, that obese individuals with increased muscle mass can maintain normal concentrations of amino acids in their blood even after more than sixty days of fasting. This is a significant finding, for we know that loss of muscle mass can be life threatening, which explains the strong association between the depletion of muscle mass (and in turn overall body mass) and how long a gravely ill patient with AIDS lives. Perhaps the best studies revealing the biology of starvation were carried out by Jewish physicians in the Warsaw ghetto's two largest hospitals from February to the middle of July in 1942. Their observations suggested that death from starvation unrelated to critical illness happens when the body doesn't have enough muscle protein to break down and furnish life-sustaining glucose, which can be manufactured in the liver from noncarbohydrate materials such as amino acids through the process of gluconeogenesis. The results of this landmark study have become a centerpiece in medical literature on the changes undergone by the human body when not enough food is available. The extensive work by Ancel Keys, a renowned scientist who studied the effects of diet (in particular, the different types of dietary fat) on health in the mid-twentieth century, also concluded that the ultimate cause of death in human starvation is total muscle mass depletion.

Just as muscle mass plays a key role in recovery from illness or trauma, muscle strength and function is critical to the recovery process. Studies have shown that muscle strength can be a factor in how long a person takes to recover from a serious illness. The less muscle mass and strength before the illness or injury, the longer it takes to heal and get back to a normal life. This is why a significant loss of muscle mass during the recovery process may push an individual who was already muscle deprived over a threshold that makes recovery of normal function unlikely to ever occur. Given this, it's no surprise that a substantial number of older women who break a hip in a fall never walk again.

It's increasingly understood that chronic diseases related to poor lifestyle account for more than two-thirds of deaths in the US. It's also becoming increasingly accepted that alterations in muscle play an important role in the most common ailments. Look no further than the ravages of heart disease and cancer to comprehend the connection. These are the major ailments suffered in the country, both of which are often associated with rapid and ample loss of muscle mass and metabolic function (a condition referred to as *cachexia*, characterized by a wasting away of the body). Survival in patients with cachexia due to cardiac conditions and cancer can often hinge on how much muscle mass they lose.

Gradual muscle loss over time that goes with aging, which accelerates the older one gets, is also related to how long one actually lives. *Sarcopenia* is the medical term for the progressive loss of muscle mass and function that typically occurs with aging; in extreme cases, one's quality of life can be decimated. Imagine not being able to perform basic daily activities, much less rise out of bed unaided, walk, or use and move your body to take care of yourself via eating, bathing, dressing, grooming, homemaking, and engaging in leisure activities. That's the impact that a devastating loss of muscle mass inflicts. Suffice it to say muscle mass and strength are key to survival, arguably as fundamental as oxygen and food.

The Fit-Fat Paradox Shapes the Obesity Paradox

As the studies about the benefits of fitness despite the presence of fatness (a "fit-fat" paradox) were coming into view, so, too, were hints that obesity as an absolute evil condition begged to be reconsidered. As a prelude to our detailed discussion of the obesity paradox, let me showcase a relatively recent study that's tested conventional health wisdom for its surprising, unassailable evidence that obesity isn't always bad.

In 2012, a report from the Swedish Coronary Angiography and Angioplasty Registry made international headlines with data from more than sixty-four thousand patients that featured a most counterintuitive and unlikely relationship: Once people have developed heart disease, they have a reduced risk of dying if they are overweight or obese, while underweight and normal-weight patients have an increased risk. The researchers evaluated patients who had developed heart problems such as unstable angina (chest pain caused by an inadequate blood supply to the heart) and heart attacks, and at those who underwent coronary angiography, a special X-ray test to discover detailed information about the condition of their coronary arteries, between May 2005 and December 2008. Patients who were underweight (BMI of less than 18.5) had the greatest risk of dying—double that of normal-weight patients, who had BMIs between 21 and 23.5. Compared to the group with lowest risk of mortality, meaning those with BMIs of 26.5 to 28, underweight individuals had three times the risk of death. The researchers also found that the relationship between BMI and mortality was U-shaped. In other words, those with the lowest risk of death were overweight and obese patients, with BMIs ranging from 26.5 to about 35; and the highest risk resided among underweight and morbidly obese patients, those with a BMI above 40.

While many of us struggle daily with constant thoughts about losing weight (and how we're going to do that), especially if we've been diagnosed with a "weight-related" condition such as heart disease, these researchers dared to suggest that there's no evidence to prove that weight loss helps someone who has a heart condition. They were even bold enough to say that there's evidence to show that weight loss after the diagnosis of a heart condition might in fact have a negative effect, particularly for those who have a BMI less than 40. I personally feel that purposeful weight loss, when done in a healthy manner and with good intentions, is safe despite the obesity paradox (see chapter 10 for my "prescription" on weight loss efforts). There is some evidence from small studies that show such weight loss can be safe and beneficial, but we have yet to collect substantial data from large studies proving the benefits of weight loss on cardiac patients and in terms of overall survival, especially in people with established heart disease.

So if you have to choose between being fit or fat, go for fit even if it means being heavier. As I've already mentioned, many people lose fitness and gain fatness with aging. The good news is that if you gain weight as you age, your health can still be excellent so long as you remain fit. And we're not talking about super-high levels of fitness, either. You just need to avoid being in the bottom twentieth percentile for fitness given your age and gender, as chapter 10 outlines. You also need to avoid the dreadful act of prolonged sitting, which could potentially cancel out much of the beneficial effects of a single daily dose of exercise due to sitting's negative metabolic outcomes. Remember, although genetics also have an influence, fitness is achieved primarily through regular, intermittent physical activity that gets your blood moving at a faster clip. In fact, many heavier people with more body fat who remain as active as possible throughout the day can be fitter—and healthier—than thin people who sit all day and do not exercise or who confine their workouts to a single time period.

And indeed, obesity does have its survivalist advantages. It is said that many great leaps in science have been made when alert people, looking for something else, came across a major discovery and recognized its importance. The discovery of the obesity paradox is turning out to be a fine example.

{ CHAPTER 5 }

Getting to the "Fat" of the Matter: The Obesity Paradox

In life we run into contradictions all the time. Jumbo shrimp. Wise fool. Bittersweet. And when there's a declaration, opinion, or proposition that seems to contradict itself or is otherwise illogical, we call that a paradox. It often contains two assertions that are true but cannot both be true at the same time. "You should not go near the water until you learn to swim" is a paradox, since you clearly can't acquire the skill until you get in the water. Likewise, in science, a paradox is any conclusion that at first sounds absurd due to established understandings about a particular set of circumstances, but that nonetheless has a viable sustaining argument. When studies have unexpected results, it's our duty to go further with inquisitive eyes and ears—to find logic in the illogical. A good paradox in science is an excellent problem to have if you're looking for the truth; it opens the door to new information, challenging long-held assumptions and compelling scientists to revisit definitions carved in reference books. It's equal parts terrifying and exhilarating.

This brings us to the obesity paradox: While obesity, particularly the severe type, is associated with higher mortality rates in the general population, modest overweightness or mild obesity can confer a survival advantage and increase our longevity. As we've started to discuss, in study after study, overweight and moderately obese patients with certain chronic diseases often live *longer* and fare *better* than normal-weight patients with the same ailments. How can this be possible?

One of the most famous paradoxes in recent memory to emerge in science, which has since helped us to embrace such anomalies, is the French paradox. In the 1980s, French epidemiologists (who study patterns of disease in populations) noticed that their fellow countrymen and women boasted low incidences of coronary heart disease despite their high intake of cholesterol and saturated fat. Given what we thought we knew about the relationship between consuming a high-fat diet and the risk for heart disease, this didn't make sense. It was practically heresy to suggest that dietary fat had nothing to do with heart disease. A typical French person eats more than forty pounds of cheese a year, and a glance at any lunch—pig's head pâté, potatoes in oil, double-fat sliced tripe, and hot sausage—would make most health-conscious Americans queasy. Despite their fatty indulgences at meals, many of the French continue to enjoy relatively good health (and, more often than not, enviable waistlines). They are famous for having lower rates of heart disease and obesity than Americans. Something about their diets is protecting them from the fate many of us experience when we consume too much artery-clogging fat. No doubt something about their lifestyle in general is offering protection, too, for many people in France and Europe probably do more physical activity (i.e., more walking to use public transport, more cycling as opposed to driving a car) than the typical American.

The French paradox quickly became a subject of intrigue among scientists around the world, and today numerous theories exist to help explain it. Not only has it shown that the hypothesis linking saturated fats to coronary heart disease is not completely valid, but it has also brought to light the impact of certain dietary habits that affect one's risk for heart disease. The French keep close tabs on their portions, and they love their wine, especially red, which is known to have a flushing effect in the bloodstream by removing blood platelets, essentially preventing those cells from clinging to arterial walls and creating deadly plaques.

The French paradox poked alarming holes in our collective medical knowledge, and it continues to be a popular field of inquiry because it has so many implications for anyone who wants to avoid heart disease. My guess is the obesity paradox will likewise shift how we perceive body fat and the factors that contribute to health, pointing as it does to an inverse relationship between body fat and risk of death in many cases.

While it's well established that obesity has contributed to our challenges with chronic illness, and that exercise is powerful medicine, no one has explained some of the unexpexted reactions taking place deep within us that have everything to do with fat's positive effects. And as I've been discussing, new science is twisting the arm of our medical community, raising provocative questions: Can being overweight or even somewhat obese actually be protective against mortality? More to the point: Are the concerns about overweight as currently defined unfounded?

In the French paradox, red wine seems to be one of the magic potions for preventing heart disease despite the prevalence of saturated fat in the French diet. The other piece to the magic is likely more physical activity and exercise; in the obesity paradox, cardiorespiratory fitness and muscle mass seem to work as the dynamic duo, helping stave off all the ills that excess body fat can otherwise have.

The Story of the Obesity Paradox

As you know by now, obesity is the second leading cause of preventable death in the US after smoking. We may see a decline in our life expectancy due to the increasing prevalence of obesity and, especially, more severe obesity, and it may soon cause more deaths per year than nicotine. It's also an independent risk factor for cardiovascular disease, meaning that obesity alone—regardless of other factors like age or cholesterol level—will increase your chances of having heart problems. But here's the problem with this gloomy obesity picture: Obese patients with established cardiovascular disease experience a prognosis that is generally equal to or even superior to those for leaner patients.

I first started to document this paradox in 2002, when analyzing several patients with heart failure who were overweight or obese but nonetheless living longer than thinner equals with the same heart problems. It didn't make sense: If obesity was bad for heart health, then how could it be beneficial in surviving heart disease? There had been some recently published studies that suggested that in patients with chronic systolic heart failure (a type of heart disease whereby the heart muscle cannot pump the blood out very well), obesity was associated with better survival. But those studies had failed to evaluate more comprehensive body composition parameters, such as percentage of body fat, body mass index, total fat, body surface area, and lean body weight (i.e., fat-free mass). So I set out to perform my own investigation with colleagues at the Ochsner Heart and Vascular Institute, including Drs. Richard Milani and Hector Ventura, Ahmed Osman (a former cardiology fellow), and Mandeep Mehra, a former Ochsner colleague who is now director of cardiovascular services and heart failure at Harvard. We looked back at patient records collected between January 1996 and December 1998 on 209 people who had been referred to our cardiac rehabilitation program to do a cardiopulmonary stress test. All of them had been diagnosed with heart failure and were taking medications to stabilize their condition. (They were at the clinic to undergo cardiopulmonary stress tests.)

Because they were simultaneously tested for all the various types of body composition parameters during their treatment, they were an ideal group to study. We compared the 28 patients who had major clinical events during that time period (i.e., 13 cardiovascular deaths and 15 urgent heart transplantations) with 181 patients who did not have such occurrences. We also evaluated each body composition parameter with regard to these major events and analyzed characteristics of obese (BMI greater to or equal to 30) versus nonobese (BMI less than 30) versus lean (BMI less than 25) patients. And lo and behold, we came across the same bewildering incongruity: an inverse relationship between obesity and body composition and subsequent clinical outcomes in patients with this common type of heart failure. The individuals who lived the longest were the fattest. We were even able to put a percentage on the risk for death in relation to body fat: For every 1 percent increase in body fat, there was a 13 percent *increase in survival.* In our conclusion, we were very cautious about our findings, stating that we didn't know if the link we'd identified was causal or merely an association. We called for more studies to confirm the connection, figure out if extra body fat was the real reason behind a better survival rate, and identify the underlying mechanism.

These findings created an uproar among my colleagues. Unfortunately, the science world wasn't ready to accept this observation in the early 2000s. The reviewers at the major medical journals resisted, despite what the research showed (one even suggested that there was a fatal flaw in our data). My initial study was finally published in 2003 by *The American Journal of Cardiology*, and it has now been heavily cited by scientists around the world. Since then, hundreds of other scientific papers have been published by major journals documenting this paradox.

Although the obesity paradox has been discounted by some experts, who have suggested that it may be due to such factors as sample size errors or variables that weren't tracked and measured (such as smoking habits), even very large meta-analyses that compile multiple studies have demonstrated this biological paradox in heart patients, most of them correcting for smoking and some even correcting for smoking-induced lung diseases. In fact, when one prominent group of researchers at the Mayo Clinic, led by my friend Abel Romero-Corral, evaluated forty studies of more than 250,000 patients with coronary heart disease, they demonstrated that in patients grouped according to BMI, those in the lowest or "normal" BMI group had the highest all-cause mortality (i.e., they were most at risk of dying from any condition or illness), whereas better survival was observed in higher-BMI groups. The overweight individuals had the lowest relative risk, whereas obesity and severe obesity had no increased mortality risk. These results were published in *The Lancet* in 2006. Two years later, a similar study by a friend and colleague Dr. Antigone Oreopolous at the University of Alberta, evaluated twenty-nine thousand patients from nine major heart failure studies and corroborated the same pattern: The overweight and obese folks with heart disease were less likely to die than the normal-weight patients with heart disease. And then in 2009, a look at eleven thousand Canadians over more than a decade found more indisputable proof: Those who were overweight had the lowest chance of dying from any cause.

Still, the critics love to point to other possible flaws in the "inconvenient research," looking for the excuse to keep fat permanently on the con ledger. Many have blamed the obesity paradox on relatively poor accuracy of BMI to reflect true body fatness. Although this measurement system does have its shortfalls, the research community has accounted for them, and has continued to find the same outcomes. In my own research, I've shown that body fat in both patients with heart failure and coronary heart disease, especially when combined with fitness, is an independent predictor of better survival. So while extra fat may contribute to the onset of certain chronic conditions, it can help you live longer with these conditions. Sounds like a contradiction, but that's the core essence of the paradox.

As I noted in the first chapter, the most compelling findings, giving indisputable credibility to the obesity paradox, have nothing to do with heart disease and everything to do with other conditions. The obesity paradox has been documented in a host of other chronic ailments, such as arthritis, kidney disease, diabetes, cancer, and even HIV infection. Although we may assume that being overweight with these conditions will make them all the more challenging to manage, treat, and combat, the evidence reveals the contrary: People with these ailments who are overweight or even mildly obese fare better in the long run than their normal-weight counterparts. Go figure.

When we first documented the obesity paradox, we initially noted the logical facts: Overweight and obese people have more cardiovascular risk factors (e.g., high blood pressure, high blood fats, high blood sugar, high markers of inflammation), and as such, they develop more cardiovascular disease, including hypertension, coronary heart disease, heart failure, and atrial fibrillation. But once heart disease is present, the overweight and the mildly obese do better than patients with normal BMIs. My research has further shown that the obesity paradox is especially apparent among unfit individuals. Unfit, slender folks with heart conditions fare much worse than their fatter but fitter peers, who have a higher BMI, higher body fat percentage, and higher waist circumference. On the other hand, fit individuals (defined by not being among the bottom 20 percent of the fitness curve for their age and gender) have a favorable prognosis despite excess weight.

The Bigger Picture

As someone who studies the heart, I am moved by these discoveries; while they defy certain age-old beliefs in medicine, they give incredible insight into the inner workings of the human body. And I have been most impressed by the associations made between the obesity paradox and other fields of medicine, some of which are related to the heart and others that center on conditions far from it. When you're searching for answers in one area of medicine, such as cardiology, and you manage to unexpectedly stumble upon clues to understanding the human body as a whole that inform virtually every facet of medicine, that's pretty remarkable. Let's summarize some of these key discoveries made within and beyond heart medicine that add to the growing body of evidence supporting the existence of the paradox.

Dialysis patients: As early as 1982, obesity was linked to better survival in dialysis patients—people with chronic kidney disease who need to be hooked up routinely to a (dialysis) machine that removes waste and excess water from their blood because their kidneys are no longer up to the task. (Chronic kidney disease is connected to the heart, as cardiovascular disease accounts for 40 percent to 50 percent of deaths in patients with end-stage kidney disease.)

In 1999, Dr. Erwin Fleischmann and colleagues at the University of Mississippi described the obesity paradox in a group of 1,346 primarily African-American men and women undergoing dialysis (though they didn't use the term *obesity paradox*). The researchers reported that the chance of dying within a year was 30 percent *lower* with each one-unit increase in body mass index over 27.5. Similar findings were reported by University of Michigan researchers in a 2001 group of 9,714 dialysis patients (of mixed ethnicities).

In a 2006 study led by my friend and colleague Dr. Kamyar Kalantar-Zadeh, of the David Geffen School of Medicine at UCLA, his team reported that among dialysis patients, obese patients are far more likely to survive than thinner patients. This study concluded that as dialysis patients tend to suffer from malnutrition and inflammation, obesity helps to counteract these problems. The excess fat provides extra calories and reserves that lower their risk of death. These studies reflect just a small slice of the evidence we now have to conclude that, generally speaking, heavier dialysis patients have a lower chance of dying than those who are of normal weight or underweight.

Diabetics: In 2012, diabetes researcher Dr. Mercedes Carnethon, of the Northwestern University Feinberg School of Medicine, published a study in *The Journal of the American Medical Association* that revealed her troubling observation: She counted sizable numbers of normal-weight people who were developing type 2 diabetes. Carnethon and her team discovered this when they examined data from five previous studies that were following people for heart disease risk factors. The studies were performed between 1990 and 2011 and included 2,625 people who were recently diagnosed with diabetes. About 12 percent of these individuals were at normal weight but were metabolically similar to those who were obese. And the kicker was that the normal-weight folks whose bodies acted obese from a metabolic standpoint were *twice as likely to die* at any point as their heavier counterparts.

The death rates remained the same even after Carnethon and her colleagues factored out other potential risk factors known to contribute to diabetes, such as high blood pressure, high cholesterol levels, and smoking, and controlling for the possibility that the normal-weight patients were losing weight due to other underlying illnesses. Overall, the death rate in overweight and obese people with diabetes was 1.5 percent per year, compared to 2.8 percent in thin patients.

The elderly: The prevalence of overweight and obesity in the elderly has been a growing concern among doctors for some time now. But recent evidence indicates that in the elderly, obesity is associated with a lower—not higher—mortality risk. Adding to research that's been accumulating over the past decade, in 2012 researchers at Tel Aviv University found that overweight people older than eighty-five might live longer than their normal-weight equals. They went on to suggest that obesity might protect seniors' bones from falls or fractures and provide energy reserves in times of stress. Proof of this counterintuitive relationship between obesity and mortality among the elderly is now strong enough that scientists in this area are recommending that doctors ditch pushing weight loss and instead encourage their older patients to prevent functional decline and muscle loss through increased physical activity, including resistance training. These scientists are also hoping that the definition of obesity in the elderly will change, since classic BMI metrics don't seem to be accurate in this population.

Rheumatoid arthritis patients: In rheumatoid arthritis (RA), an autoimmune disorder in which the body attacks its own tissues and joints, a near-linear relationship exists between increasing BMI and lower mortality, with the effect existing even in severe obesity. Although higher weight may have some adverse effects on the arthritis itself (e.g., weight puts mechanical stress on the back, hips, knees, ankles, etc.), it appears to at least be associated with better survival in RA. In 2005, researchers at the University of Texas Health Science Center at San Antonio published a study in the *Archives of Internal Medicine* featuring a prospective group of 779 RA patients. Their mortality rate over 4.4 years of those with a BMI greater than 30 was 66 percent *lower* than that in patients with a BMI of 20 to 24.9, even after adjustment for smoking, duration of disease, and medication.

Now, let's take a moment to connect some dots here, because they may not seem obvious. Interestingly, RA severely increases one's chance of having a heart attack—within the first ten years of diagnosis, one's risk is nearly doubled. This is primarily caused by the raging inflammation that arthritis causes throughout the body, which impacts the heart too.

On the surface, it might not seem logical that obesity offers protection against earlier death in those with severe RA. As we're about to see with other types of conditions known to cause a general wasting away of the body, arthritis of any kind causes destruction of joint cartilage and bone and is accompanied by a loss of body cell mass, known in medical circles as *rheumatoid cachexia*. This condition is present mostly in skeletal muscle, but also occurs in the internal organs and immune system, leading to muscle weakness, a loss of functional capacity, and potential acceleration of death. And therein lies the protective value of body fat: It can provide vital reserves that slow the body's deterioration under the duress of the arthritis. Put simply, it acts as a much-needed shield—and cushion—against the complications that emerge as a result of all this muscle and tissue loss in the wake of extreme arthritis.

People with illnesses that cause uncontrolled, dangerous weight loss: Anyone who has witnessed someone die of a disease like cancer or an AIDS-related illness knows the physical toll such afflictions often take: The person languishes over time, losing vast amounts of weight and dying bone-thin. The Seven Countries Study, which has followed thirteen thousand men over the past forty years, is the first to systematically examine the relationships between lifestyle, diet, coronary heart disease, and stroke in different populations from different regions of the world, and has found that the risks of dying from cancer and infections decrease with *increasing* weight. This makes sense when you think about it: If you're overweight and diagnosed with a fatal illness, you have extra reserves to help fight the disease for as long as biologically possible.

Beliefs about the ramifications of obesity have become so widespread that few people stop to question if these seemingly universal adverse effects have been categorically proven. But when science is faced with statistics and data like those described in the previous pages, we have to ask why. Although all the existing research and comprehensive analyses indisputably demonstrate an obesity paradox, we are still in the beginning stages of figuring out why it exists.

The Fa(c)ts of the Matter

Before we get into the details of the four main explanations for the obesity paradox, I want to remind you about two facts that should always be kept in mind. One is that fat helps guard the body from damage, particularly as we age and are more prone to injury and falls. It literally provides a physical barrier against traumatic injuries. Second is that the body mass index formula is, as you've learned, flawed. These two facts alone help explain the obesity paradox from a very general standpoint. But, as they say in infomercials, wait—there's more.

The Power of Extra Reserves

Anytime the body is fighting an illness or shouldering the weight of a chronic disease, it requires more energy and higher caloric reserves than usual, so it makes sense that extra weight in the form of body fat is helpful. If you're someone with a chronic condition and you lack those reinforcements via extra fat, you may become malnourished even though your weight is considered normal. Although body fat releases a plethora of bad molecules and hormones that antagonize health, it also secretes certain *beneficial* substances that can aid in a fight against illness. It's well documented, for instance, that fat tissue and fatty molecules circulating in the blood play a role in reducing some of the deleterious effects seen in chronic diseases. So the more body fat you have, the more ammunition you have in your arsenal to combat sickness and keep those troops well fed.

Remember, too, that location matters. Oxford University researchers, for example, have found that adults with pockets of lower-body fat in the buttocks and thighs might have a reduced risk of diabetes and heart disease because this fat traps the potentially harmful fatty acids that can travel through the bloodstream (and to the heart).

The Power of Genetics . . . and Your Environment

We can't deny the impact that certain gene variants can have on people to make them more (or less) susceptible to illness, as well as put them at much greater risk of death once they become ill. Seemingly healthy, slim people who develop diabetes, cardiovascular disease, or another chronic ailment could simply have their genes to blame. And heart disease in thin people may very well represent an entirely different illness from heart disease in heavier folks. Unfortunately, thin patients often are not treated as aggressively because there's a built-in bias toward thinking they are healthy. By the same token, overweight and obese individuals may receive earlier and more proactive care as a result of their obvious condition (What's more, since overweight and obese people have higher levels of blood pressure, they may tolerate higher doses of cardio-protective medications used to treat patients with heart disease.) So some of the obese people who fare better than their thinner comrades could attribute their prognosis to superior genes, as well as more attention to their health in general. Additionally, as I explained earlier, the obese patient with heart disease may not have developed the disease in the first place without weight gain during his or her adult life, whereas the thin person with the same heart disease may be doomed by a bad gene pool.

That said, we cannot forget that the more we learn about our genetic makeup, the more we realize that our environment (which here refers to both environmental factors and lifestyle) plays a much bigger role in how our DNA expresses itself than we ever imagined. Whether or not we fulfill the prophecy of some genetic predisposition to develop heart disease or diabetes is largely up to our lifestyle choices turning genetic switches on or off. Genes are seldom the sole predictors of our destiny. The environment, which combines overlapping influences from diet and exercise to exposure to toxins and stress, can both directly and indirectly impact our health risks. It can ultimately affect the genes that we've inherited for good or bad. Bear in mind that on the genetic side of the equation, we're talking inherited *risk factors*—not necessarily guaranteed triggers of disease. This is an important distinction to remember. Far too many people stand by fatalistic views when it comes to their DNA and how it impacts their health. You can take control of the environmental side of the equation and considerably reduce your overall lifetime risk of developing most diseases. And as I've shown, staying fit can contribute immensely to physical health, but it can enhance genetic health too, affecting longevity. Dietary choices and exercise are two of the biggest environmental factors that influence the state of our metabolic health and whether or not our body fat helps or hurts us.

The Power of Fitness

Much of what we know about the link between obesity and health has been derived from research like the Framingham Heart Study, which has followed nearly fifteen thousand men and women since the 1940s. But sometimes Framingham and other longitudinal studies have failed to take into account physical activity and fitness. Only in the past several years have we been able to distinguish weight and fitness in terms of health. While the importance of dietary factors continues to be squarely placed in the picture of health, we now have a new piece of data to consider. Indeed, it's much better for your health if you're fat and fit than thin and unfit. As chapter 4 conveyed, your fitness level—encompassing both muscle mass and cardiorespiratory fitness—is a far more important predictor of your risk of dying than just knowing what you weigh.

The impact of fitness on genetic expression cannot be understated. As I outlined in the previous chapter, this is where some of the most provocative research has taken place lately, as scientists work to demonstrate how regular exercise can turn on our "youth genes" and keep potentially harmful genes from blinking on. In 2011, Dr. Mark Tarnopolsky, a professor of pediatrics at McMaster University, in Hamilton, Ontario, discovered that exercise kept a strain of mice from becoming prematurely gray, and altered their course of aging by stimulating the production of certain molecules in the body that have far-reaching effects on the body's natural antiaging mechanisms, including repairing damaged DNA and preserving mitochondrial health.

Fitness could very well be the single most important ingredient to living as long as possible, despite all the other risk factors we bear. In my own research, I've witnessed time and time again that despite people's high percentage of body fat and chronic conditions, if they are fit, they are healthy by many standards and will outlive their thinner comrades who don't ever sweat.

The (Unfortunate) Power of Bias

Are we asking the right questions and approaching the right problems in the first place? Or are we falling prey to inherent biases that color our viewpoints and handicap our ability to distinguish between fact and fiction? I think it's fair to say that we've gotten so used to framing health issues in terms of obesity that we could be overlooking other potential causes of disease. We assume that obesity—and its underlying content, fat—is inherently and invariably bad. But if we're less judgmental when we consider the data, we can often find confounding factors that can explain the disease associations we blame on weight. We covered some of these issues in chapter 2 when we discussed "obesity-related" conditions and illnesses, all of which can occur in the absence of overweight and obesity.

In 1981 Dr. Neil Ruderman, an endocrinologist at Boston University School of Medicine, was among the first to take note of people with "normal" BMI numbers who also had metabolic abnormalities, including high levels of insulin resistance and blood fats (triglycerides); they tended to carry belly fat, which, as I've already covered, is more likely to affect vital organs than fat in the hips and thighs. And then we have the people who comprise the most intriguing category of all, Metabolically Healthy Obesity: the individuals who are obese but who don't have any of the conditions we often associate with obesity such as hypertension, high blood lipids, high blood sugar or diabetes, and signs of chronic inflammation. Why has it taken so long for science to rectify these contradictions? We blame obesity for most everything these days, yet have a hard time making exceptions to the rule. We'll meet a man in the next part whose "metabolically obese normal-weight" condition points to the true killer we should be blaming: prolonged blood sugar imbalances over time, which can exist regardless of weight.

So Where's the Sweet Spot?

It's a question that I find many ask once they've started to digest some of the characteristics of the obesity paradox. If fat has a good side, then what is the ideal amount to have if we want to avoid not just heart disease but any life-threatening condition? Let's turn to two revealing studies, one of which I mentioned in chapter 1 and was published in *JAMA* around New Year's in 2013 (fitting for diet season). Because it suggested that the ideal body mass index to have is between 25 and 30—not the 18.5 to 24.9 that we're used to accepting as "ideal" or "normal"—the study was met with scorn in the scientific community. But it really began in 2005, when the study's lead author, epidemiologist Katherine Flegal, analyzed data from the National Health and Nutrition Examination Survey (known as just NHANES, a six-year survey of more than thirty-three thousand Americans from two-month-old babies to seniors over seventy-five years). She found that the biggest risks of death were associated with being at the far edges, either underweight or severely obese. The lowest mortality risks were among those in the overweight category (BMIs of 25 to 30), and being moderately obese (30 to 35) was no more risky than being in the normal-weight category.

Flash forward a few years. Dr. Flegal and her colleagues at the Centers for Disease Control and Prevention's National Center for Health Statistics decided to confirm the NHANES results further when they reviewed ninety-seven separate studies, encompassing a staggering 2.9 million people. After all, when one study proves a point, it helps to keep proving it again and again with even bigger populations. And they arrived at the same conclusion: People who were overweight (BMI between 25 and 30) were often better off than those in the moderately (BMI above 35) or morbidly obese (BMI above 40) or normal-weight categories. And the mildly obese (BMI between 30 and 35) were not positively or negatively affected (they actually had a 5 percent lower risk of dying, which is not quite significant by standard statistical methods). To put this research into crisp perspective, the numbers boil down to this fact: Being overweight is linked to a statistically significant 6 percent *lower* risk of dying, compared to people considered to have an ideal weight. Now, as I'll clearly spell out in chapter 10 when I offer my prescriptions, I am not suggesting that someone with a BMI in the low 20s should try to gain weight in order to get into the upper 20s; nevertheless, the current large-scale research suggests that one can have a quite healthy weight, at least with regard to overall survival, in the upper 20s for BMI (or even with a BMI in the low 30s), and that is without even factoring in the very important component of physical fitness.

"Science Is Truth Found Out"

Although the linear association between increases in weight and mortality became dogma in the 1960s when the Metropolitan Life Insurance Company came out with its chart of "desirable weights," hints of exceptions to the rule would eventually become too apparent to ignore, compelling astute observers to take note. Reubin Andres was one such observer in the early 1980s, when he was the director of the National Institute on Aging, in Bethesda, Maryland. After reexamining various research studies alongside actuarial tables, he questioned the now established wisdom about the linear relationship between weight and mortality by showing that it followed a U-shaped curve when height was also factored into the equation. He also demonstrated that the bottom of the curve, the weight at which people had the lowest death rates, depended on age. According to his calculation, MetLife's weight recommendations could not be seen as black-and-white rules for everyone. Andres said that they might work for the middle-aged, but not for older folks. He stated that the data proved older individuals were largely better off if they were overweight—an aggressive statement for the decade. It was our first sign of the obesity paradox coming into view, albeit a clue that was vehemently rejected by the medical community. Andres didn't make friends over his unconventional ideas.

Plenty of opponents of Andres's viewpoint rallied against him. In a frequently cited *JAMA* paper published in 1987, for example, Walter Willett and JoAnn Manson, two leading nutrition and epidemiology researchers at Harvard, tried to throw a wrench into Andres's argument. Looking at twenty-five studies that analyzed the relationship between weight and death, Willett and Manson found that most could easily be discredited by two factors: smoking and sickness. They highlighted the fact that smokers tend to be leaner and die earlier than nonsmokers (true), and that illness, especially the chronic kind, tends to cause people to lose weight (true). So they argued that these effects were essentially shams—they mistakenly made thinness itself seem to be a risk.

One of the major hurdles in assessing the dangers of fat is that cause and effect can easily become all mixed up, a critical concept we'll be exploring in the next part. Your weight obviously factors your health, but your health can also have a huge bearing on your weight. If you have cancer or some other serious illness, for instance, it will probably cause you to lose weight. Similarly, being a smoker will make you less healthy and at the same time might make you slimmer.

Flegal's early studies on the obesity paradox, in 2005, validated Andres's U-shaped mortality curve. She not only showed that overweight people had a lower mortality rate than those of normal weight, but she further showed that the pattern remained true even among those who had never smoked.

The rift between researchers only continued to grow . . . and grow more contentious. This partly inspired Flegal to carry out her much larger and more comprehensive analysis, which shook everyone to the core with its explosive conclusions and breadth in scope. Her expansive analysis encompassed all prospective studies that investigated death rates using standard BMI categories, which amounted to ninety-seven studies in total. And she made sure that these studies did indeed account for factors that could potentially skew authentic results, such as smoking, age, and gender. The resulting combined data rang loud and clear: People whose BMIs were in the overweight range (between 25 and 29.9) showed the lowest mortality rates.

As all good debates in science go, Flegal didn't have the last word. Willett's Harvard group shot back, alleging that she failed to fully correct for age, illness-related weight loss, and especially for smoking, since there's a difference between heavy and occasional smokers (hence his insistence that the only way to correct for the effects of smoking was to focus solely on people who had never smoked rather than employ standard statistical adjustments). Willett even pointed to one of his own studies, published in 2010, that Flegal didn't include because it didn't use standard BMI categories. Analyzing data from 1.46 million people, Willett and his colleagues had found that among nonsmokers, those in the "normal" BMI range of 20 to 25 showed the lowest mortality rates.

The debate roared on. Flegal then pointed another critical finger at Willett's study for its deletion of certain data that could have changed his desired results. Willet had removed data involving nearly nine hundred thousand people, and according to Flegal, once you delete such a large number of people, you can't know the difference between those who'd never smoked and everyone else in the study. Indeed, this is a good point, for people who've never smoked in their life could share certain characteristics, such as being more educated or richer, that impact the results regardless of the potential effects of smoking. Flegal was also critical of Willett's study for another legitimate reason: He relied on people's self-declared heights and weights (creating the likelihood of inaccuracies), rather than objective measures.

Casting such scientific bickering aside for a moment, we must acknowledge that there's a deep philosophical difference at play here: our entrenched ways of thinking and subconscious biases. The belief is ingrained in our culture that those who go through the hard work of watching their waistlines and achieving thinness through diet and exercise should be rewarded with good health and a long life. As the journal *Nature* so cogently put it, "It seems so unfair to think that dumb luck plays a big role—that the genes we're stuck with determine whether we get nice subcutaneous fat that makes us look shapely or at least cuddly, or dangerous 'visceral' fat that surrounds our vital organs with a toxic, inflammatory envelope."

Additionally, different scientists, despite their best intentions, approach their studies with different goals in mind. For someone like Flegal, it's all about the data and reporting what's true despite its implications from a general public health standpoint. She's not a policy maker. Her job is simply to provide accurate information that can then be used by the policy makers and others. But for someone like Willett, the messages that the public can interpret from the data are a concern. He worries that people could take the essence of the obesity paradox as being a general endorsement of obesity. He's also concerned that all the mixed messages could diminish people's trust in science—one day we're told that obesity is bad, and then the next, it's good. Where's the truth? It's buried in a heavily nuanced ream of data that only scientists can mine and really understand.

But if we zoom out, away from the nitty-gritty details of this somewhat elitist debate, perhaps we can find merit in both perspectives, for Willett and Flegal are simply looking at data through slightly different albeit equally respectable lenses. At the end of the day, we have enough studies to prove the existence of the obesity paradox. All good scientists know that you can't argue with good data. They would also admit that it's more critical to pay attention to overwhelming data than to try and un-explain them. What people like Willett and Flegal are trying to do is explain why the paradox exists at all. After all, there are plenty of studies now to irrefutably show that in people with serious illnesses such as heart disease and type 2 diabetes, those who are overweight have the lowest death rates. Counterintuitive, yes, but a clear fact we all need to take seriously despite ingrained negative thoughts about weight. And we must acknowledge that metabolic reserves could be more important as we age than we previously thought.

Moreover, it should be recognized that the Flegal analysis just assessed mortality—a pretty hard endpoint, being dead or alive—and did not study the impact of obesity on other chronic diseases and quality of life.

There's no question that survival is a balance of risks. When you're young and healthy, becoming obese will certainly cause problems in a matter of years. With age, however, the balance may tip in favor of extra weight. And of course we cannot deny the fact that genetic and metabolic factors may also be at play despite weight, a fact I've covered throughout the book. Look no further than the astounding number of people who develop type 2 diabetes when they are of normal weight, and who are twice as likely to die over a given period as those who are overweight or obese. They are thin yet metabolically obese: They have high levels of insulin and triglycerides in their blood, both of which are strong risk factors for developing "obesity-related" conditions such as diabetes and heart disease.

All this further makes the case that BMI is a terribly unsophisticated and unreliable measure for evaluating the health of individuals. It doesn't tell the whole story. In a more perfect medical world, which I hope to see in my lifetime, we'll have more accurate and practical tools for assessing individuals and their risk for illness and premature death. As medicine advances, doctors like me will use other data—waist measurement, blood sugar, cholesterol, markers of inflammation, relevant hormones, and other measures—to complement the traditional BMI in determining which patients will really benefit from losing weight and how much (if any) weight a person should lose, and in identifying the fat-but-fit and metabolically obese. Clearly, knowledge about fitness would also significantly aid in this risk assessment.

Dr. Mitchell Lazar, of the Perelman School of Medicine at the University of Pennsylvania, wrote an article for the journal *Science* in 2013 to reconcile this incendiary debate over the obesity paradox: "The optimal weight that is predictive of health status and mortality is likely to be dependent on age, sex, genetics, cardiometabolic fitness, pre-existing diseases and other factors. To quote Galileo, measure what can be measured and make measurable what cannot be measured." I couldn't have put it better myself.

Whatever the prevailing explanation for the obesity paradox turns out to be, most experts agree that the data casts a new light on the role of body fat. Obviously, maintaining fitness is good and maintaining a healthy metabolism is good, and if you had to choose between fitness and thinness, it looks like it's much more important to maintain your fitness than your svelte waistline. Fitness appears to be a lot more protective than a low weight. That is a message that may take a long time to reach your family physician, however, and not many people are willing to challenge the weight conventions.

The pervasive and narrow-minded focus on weight loss irrespective of fitness in our culture is not good news. As the population ages, the number of men and women suffering from chronic conditions is expected to skyrocket. My hope is that this latest science gains traction in our general health community. In my own field, heart failure is still a diagnosis that carries a relatively poor prognosis. Fewer than half of patients survive five years after their original diagnosis, and only 25 percent are alive a decade later. About 22 percent of men and 46 percent of women will die of total heart failure within six years of surviving a heart attack. Similarly dismal statistics are found in other areas of medicine: Most people who have chronic kidney disease, for example, will worsen over time and eventually die of kidney failure.

And maybe the shift in paradigm here will come with a drastic rethinking of the elusive character standing on center stage: fat. It's time we got up close and personal with this indelible part of us. As I have said many times in this book, I am not trying to promote obesity and certainly not weight gain (except in the very underweight). But probably the number one preventable factor contributing to disease in our current Westernized world is physical inactivity and lack of exercise.

{ PART II }

The Purpose and Power of Fat

{ CHAPTER 6 }

Fat Fundamentals: Why Do We Need It Anyway?

Fat is one of the basic components that make up the structure of your body. The other elements include muscle, water, bone, and your organs. Obviously these are all necessary for normal, healthy functioning, but so is body fat. If it didn't have a good side, then we probably wouldn't have a need for it at all. Seems so obvious, but consider how much we tend to reflect negatively on body fat and often try to lose as much of it as possible.

Discovering fat cells' profound responsibilities in the body has revolutionized the study of obesity. In addition to their prominent job of disposing of dietary fat, fat cells have been shown to wield power over glucose metabolism, blood pressure, appetite, hormone production, and how your body utilizes its energy resources. Practically every week, the scientific literature unveils another function for the humble fat cell.

Building on what we learned about fat and the importance of its location in the body in chapter 3, here we'll take a sweeping look at the role of body fat and why it's so important for our survival. This conversation will dispel a lot of myths that prevail in society, as well as share a bevy of facts that have only recently come to light in scientific circles. No matter how much we benefit from the fat in our body, a veritable phobia about fat still exists. This is due to the many misconceptions that we have about it.

Fat Fundamentals

Fat is paramount to survival. No one can have a fat-free body and be healthy at the same time. Fat also is a core pillar of our nutrition: Each gram of dietary fat provides nine calories of energy for the body, compared with four calories per gram of carbohydrates and proteins. Another way of looking at this math is to say fat is the most efficient source of food energy.

In brief, we need fat for warmth and insulation, for protection and cushion, for energy, and even to think. About two-thirds of the brain is composed of fat; it provides the structural components not only of cell membranes in the brain but also of myelin, the (fatty) insulating sheath that surrounds nerve fibers throughout the body, enabling them to carry messages faster. Fat is in fact a vital ingredient of all the cell membranes throughout the body. And without a healthy cell membrane, the rest of the cell cannot operate successfully.

In addition to giving skin its smooth outer appearance (one of the more noticeable signs of extremely low body fat is the sight of bulging veins), the layer of fat just beneath the skin, subcutaneous fat, helps us regulate body temperature. This explains why thin people tend to be more sensitive to cold; overweight and obese people tend to be more sensitive to warmer temperatures.

When we talk about "fat" as being uniquely critical for the body, most of the time we're referring to fatty acids, important chemical compounds found in plants, animals, and microorganisms. In humans, fatty acids help control blood pressure and inflammation, and keep blood from clotting. They are the molecules that help with cellular development and the formation of healthy cell membranes, and they have been shown to block tumor formation in animals, as well as hinder the growth of human breast cancer cells.

It's not necessary to remember all the chemistry about fat, but it's important to understand some basics. A fatty acid typically consists of carbon atoms attached to each other in a straight chain. Hydrogen atoms are also attached to the carbon atoms along the length of the chain as well as at one end. At the other end is a carboxyl group (–COOH), and it's this carboxyl group that makes it an acid (carboxylic acid). If the carbon-to-carbon bonds are all single, the acid is said to be *saturated*; if any of the bonds are double or triple, the acid is called *unsaturated*. A few fatty acids have branched chains; others, like prostaglandins (hormone-like fatty compounds that participate in the contraction and relaxation of smooth muscle tissue), contain ring structures. Fatty acids are not found flying solo in nature; instead, they are found in combination with the alcohol glycerol in the form of triglyceride. Oleic acid is the most widely distributed fatty acid. It's abundant in some vegetable oils (e.g., olive, palm, peanut, and sunflower seed) and makes up about 46 percent of body fat.

Fatty acids that are necessary for survival but cannot be manufactured by the human body are referred to as *essential fatty acids*; we must obtain them from the diet. These molecules fulfill many life-sustaining roles. For starters, as I just noted, they assist in the health of the brain and nervous system, as well as regulating immune responses, liver function, and proper thyroid and adrenal activity. They help thin your blood, which is how they can prevent clots that lead to heart attacks and stroke. Their anti-inflammatory properties allow them to relieve symptoms of both arthritis and other autoimmune disorders. Essential fatty acids also aid in the breakdown of cholesterol (more on that shortly). And they serve a role in the vanity department: If you're deficient in these essential fatty acids, you're likely to suffer from skin problems, including dry, flaky skin and eczema, dandruff, split nails, and brittle hair.

Omega-3 and omega-6 fatty acids comprise the two biggest categories of essential fatty acids. Omega-6 essential fatty acids help the body cure skin diseases, fight cancer cells, and treat arthritis. Although there is a perception that omega-6 fats are bad due to their overabundance in the American diet and the fact they are proinflammatory, the truth is that we need them (in moderation). Most people get plenty of omega-6 fats daily, as they are readily found in meats and vegetable oils (vegetable oil represents the number one source of fat in the American diet). Interestingly, one of the most biologically critical fatty acids for humans is an omega-6 called *linoleic acid*, which is known to aid infants in their growth and development (and makes up the bulk of the ingredients in breast milk). It's also found in cow's milk, cheese, beef, and lamb, and has been shown to help with decreasing abdominal fat, lowering bad cholesterol and triglycerides, and increasing the speed of one's metabolism.

Omega-3s serve a variety of purposes within the body, although they are not as present in the standard American diet (and so we must be proactive in making sure we consume enough). First, they help your cells and organs to function properly, aid in the formation of cellular walls, and encourage the circulation of oxygen throughout the body. A lack of omega-3 essential fatty acids is known to lead to blood clots. If you seriously lack omega-3s, you might have problems with memory, a decreased sense of vision, hair and skin issues, an irregular heartbeat, and a decrease in the functioning of your immune system. Low levels have also been linked to many brain-related disorders such as mood swings, dementia, and attention deficit hyperactivity disorder (ADHD).

There are many sources of essential fatty acids that you can add to your diet to get the nutrients you need. We don't need much help in consuming omega-6s, which tend to be abundant in our diet. But you need a balance of both omega-3 and omega-6; too much omega-6 and not enough omega-3 will increase inflammation. The all-star omega-3 sources include:

- nuts
- soybeans
- walnut oil
- canola oil
- flaxseed oil
- olive oil
- cold-water fatty fish (e.g., salmon, herring, cod, flounder, tuna, arctic char, bluefish, and shrimp)

Untangling the Confusion Over Triglycerides

Most everyone has a pretty good idea about what cholesterol is these days (and the meaning of high cholesterol), but triglycerides still stump a lot of people. Unfortunately, triglycerides get a bad rap because they can be problematic in excess, just like cholesterol.

On a very basic level, triglycerides are blood fats that help enable the two-way movement of fat and blood glucose from the liver. As the name implies, triglycerides contain three molecules (*tri-*) of fatty acid and one molecule of glycerol. The amount of triglycerides in the blood is one important barometer of metabolic health, as high levels (anything above 200 mg/dL with values between 150 and 199 being borderline elevated) are associated with coronary heart disease, diabetes, and fatty liver disease (a serious condition in which toxic fat makes up more than 5 to 10 percent of the weight of your liver, often caused by too much alcohol or sugar consumption). There are many triglycerides, some of which are highly saturated while others are not. They are the main constituents of vegetable oil (typically less saturated) and animal fats (typically more saturated). They are also a major component of human skin oils. Triglycerides get into your bloodstream in one of two ways: They can come directly from the fats in your diet or they can be made inside the body from carbohydrates you consume.

Dietary Fats

There are three kinds of dietary fats, not including their kissing cousin, cholesterol. Saturated fats are the only fatty acids that raise total blood cholesterol levels and low-density lipoprotein (LDL, bad cholesterol). For this reason, they may increase your risk of cardiovascular disease and may increase your risk of type 2 diabetes. Usually solid at room temperature, they are naturally found in meats and whole dairy products like milk, cheese, butter, cream, and ice cream. Some saturated fats are also found in plant foods like tropical oils (coconut or palm kernel oil). Although saturated fats have been demonized for years, especially since we generally get plenty in our daily fare, every cell in your body requires them for survival. Very recent evidence has seriously questioned the adverse effects of saturated fats, which may actually be less deleterious than an excess of carbohydrates, especially in physically inactive individuals, and trans fats, which may be the most toxic. Saturated fats comprise 50 percent of the cellular membrane and contribute to the structure and function of your lungs, heart, bones, liver, and immune system. With the help of saturated fats, your liver can successfully process fat so it doesn't have detrimental metabolic effects, as well as protect you from the adverse effects of toxins, including alcohol and compounds in medications. The white blood cells of your immune system partly owe their ability to recognize and destroy invading germs and tumors to the fats found in butter and coconut oil. Even your endocrine system relies on saturated fatty acids to communicate the need to manufacture certain hormones, including insulin. And they help tell your brain when you are full so you can pull away from the table.

Trans fats are basically man-made synthetic fats that act like saturated fats and can raise your bad cholesterol level. They are produced when corn, soybean, or vegetable oil is made into a solid fat through a process called *hydrogenation* (hence the phrases "hydrogenated oil" and "partially hydrogenated oil" on ingredient lists), and are often found in margarine and processed foods like snacks (crackers and chips), commercially baked goods (muffins, cookies, and cakes), shortening, and many fast-food items. These probably are the most toxic with hardly any redeeming properties, which is why they have been banned by our Federal Drug Administration (FDA).

Unsaturated fats are usually liquid at room temperature. They are found in most vegetable products and oils, and are often categorized as being monounsaturated or polyunsaturated. (Monounsaturated fats have one pair of carbon molecules joined by a double bond; polyunsaturated fats, on the other hand, have two or more double bonds between carbon atoms in the carbon chain backbone of the fat.) Studies show that eating foods rich in monounsaturated fats (MUFAs) improves blood cholesterol levels, which can decrease your risk of heart disease. Research also shows that MUFAs may positively impact insulin levels and blood sugar, which can be especially helpful if you have type 2 diabetes. Polyunsaturated fats (PUFAs) are found mostly in plant-based foods and oils. Evidence shows that eating foods rich in PUFAs also improves blood cholesterol levels, and may help decrease the risk of type 2 diabetes. Omega-3 fatty acids are but one type of polyunsaturated fat, which may be particularly protective.

Technically, cholesterol isn't always considered to be a fat, although it's just as vital. Rather, it's a waxy, fatlike substance that every cell has the capacity to make. Contrary to what you might think, the two types of cholesterol we hear about, HDL (high-density lipoprotein) and LDL (low-density lipoprotein), are actually not two different kinds of cholesterol. HDL and LDL reflect two different containers for cholesterol and fats, each of which serves a unique role in the body.

In general, cholesterol forms cell membranes with other saturated fats. It also helps guard those membranes and police their permeability so different chemical reactions can take place inside and outside the cell. The gallbladder's bile salts, which are secreted to digest fat and facilitate the absorption of fat-soluble vitamins, are made of cholesterol. Having an extremely low cholesterol level in the body would therefore compromise your ability to digest fat. It would also jeopardize your body's electrolyte balance, since cholesterol helps manage that delicate equilibrium.

Cholesterol further supports brain function and development. The brain holds only 2 percent of the body's mass but contains 25 percent of its total cholesterol. That's right: One-fifth of the brain by weight is cholesterol. We've actually determined fairly recently that the ability to grow new synapses in the brain depends on the availability of cholesterol, which latches cell membranes together so that signals can easily jump across the synapse. Earlier I mentioned how fats serve an important role in cellular membranes and message transmission. Well, cholesterol has a starring role, acting as a crucial component in the myelin coating around the neuron to allow for quick broadcasting of information. What's more, cholesterol in the brain serves as a powerful antioxidant. It protects the brain against the damaging effects of free radicals. It's also a precursor for the steroid hormones like estrogen and the androgens, as well as for vitamin D, a critically important fat-soluble antioxidant. There may be a sound reason why natural cholesterol levels generally increase in the body as we age. As age brings on higher levels of free radical production, which promotes faster aging, cholesterol can offer a level of protection. In fact, there's recent evidence showing that both fat and cholesterol are severely deficient in diseased brains and that high total cholesterol levels, at least late in life, are not always associated with decreased longevity. Lest you think only the elderly are affected, we also know that depressed individuals who are underweight sometimes don't respond to antidepressant therapy because their brain's receptors, which are comprised of fatty acids, cannot function properly to transport and utilize these medications. Without enough fat, the brain can't maintain its receptors that bind to certain molecules, like those from drugs, to change brain chemistry.

Obviously, as a cardiologist, I treat a lot of patients with high cholesterol, and one of my goals in treatment is to lower their blood levels of LDL cholesterol, since too much does have its negative repercussions on the body. But I don't drive these values to dangerous low levels because the body—and especially the brain—still needs a certain level of LDL to survive. An essential, healthy value of LDL cholesterol to have is between 50 and 100 mg/dL.

In addition to helping provide structure and function to various parts of the body, one of the chief reasons for taking in fat through our diet is to help us absorb and use essential fat-soluble vitamins such as A, D, E, and K. Because these vitamins do not dissolve in water, they can only be absorbed from your small intestine in combination with fat. Without enough vitamin K, you'd lack the ability to form blood clots and could suffer from spontaneous bleeding. A deficiency in vitamin A would render you vulnerable to blindness and infections. And a lack of vitamin D is known to be associated with increased susceptibility to several chronic diseases, including depression, neurodegenerative diseases, and a number of autoimmune diseases, such as type 1 diabetes and may increase the risk of heart diseases, especially hypertension and heart enlargement. It doesn't help that the American diet is already vitamin poor due to an overabundance of processed, unnatural foods. And when we avoid eating fat we put ourselves at even more risk of suffering from vitamin deficiencies.

Fat: The Four-Star General

Of all the functions and purposes of fat in the body, it has four key responsibilities that relate directly to the obesity paradox.

Fat bolsters immunity. Cells that are precursors to fat cells, called *preadipocytes*, act like special immune cells that can devour invading germs and bacteria. This partly explains why people who diet to extremes tend to report getting sick more often and feeling more inflammation, from general aches and pains to joint issues.

Fat stores glucose. Body fat is one of the places, along with muscle and liver, where your body stockpiles glucose, storing it as fat molecules. Studies of rats and humans who have little body fat show chronically high blood glucose levels, just like in diabetics. This is because muscle tissue has a limit to how much glucose it can hold in the form of glycogen. Diabetics can't get blood glucose into their cells because they either lack sufficient insulin (type 1 diabetes) or their cells don't respond to insulin (type 2 diabetes). But without enough body fat to store extra glucose for a rainy day, that extra glucose can end up staying in the blood, where it can wreak havoc on the body. Biologically speaking, an anorexic body can actually act like a diabetic one.

Fat produces hormones. In addition to storing and releasing triglycerides, fat cells produce an enormous array of hormones pivotal to health. Because everything in the body is connected, shifts in hormones through the years can have profound effects on the body. Our hormones have as much to say about our reproductive system as they do about our hunger and our ability to gain and lose weight, maintain a healthy immune system, keep a beating heart, recover from injury, think clearly, sleep soundly, have a general sense of well-being, and so much more. To keep the body balanced, the forces of one particular hormone are usually counterbalanced by those of another hormone.

While leptin is in charge of telling your brain to stop eating once you've had enough, for instance, ghrelin is responsible for triggering feelings of hunger when you haven't eaten in a while and the body needs fuel. Similarly, insulin is our chief anabolic hormone, maintaining stores of energy and building them up after meals. On the other hand, glucagon, growth hormone, adrenaline, and noradrenaline counteract insulin's actions, making energy available in the form of fatty acids or glucose when these are needed. If you disrupt your body's natural hormonal state, you render it vulnerable to an onslaught of health problems, from minor to life threatening. Which is why we really can't say any single hormone is categorically "good" or "bad"; they only take on such labels when there's too much of one and not enough of another to maintain the balance necessary for a healthy metabolism.

Physiologically, hormones control much of what you feel—hungry, thirsty, moody, tired, sick, energetic, hot, or cold. The endocrine system lords over development, growth, reproduction, and behavior through an intricate, balanced orchestration of hormones. On a basic level, hormones are your body's messengers—typically produced in one part of the body, such as the thyroid, adrenal, or pituitary gland, they pass into the bloodstream, where they can be transported to distant organs and tissues to modify structures and functions. They essentially act like traffic signs and signals, telling your body what to do and when so it can run effectively. They are as much a part of your respiratory, cardiovascular, nervous, muscular, skeletal, immune, and digestive systems as they are a part of your reproductive system.

In addition to estrogen, the most familiar hormones include progesterone, cortisol, adrenaline, and androgens like testosterone. Every organ has certain hormones, and many hormones have multiple functions that overlap. The endocrine system collaborates with both the nervous and immune systems to help the body successfully manage different events and stressors. Just as too much or too little of certain hormones can lead to obesity, obesity can lead to changes in hormones and their levels. The following are just a few of the many hormones generated by fat:

• *Leptin:* As mentioned, this hormone is involved in dozens of bodily processes but is most known for its role in appetite suppression and controlling how the body manages its fat storages. It reduces the urge to eat by acting on specific centers of the brain. And since leptin is produced by fat, higher levels of it are usually found in obese people. Curiously, research has shown that obese people are less sensitive to leptin's effects, so they don't feel adequately full after eating. Ongoing research is trying to figure out why this happens.

• *Angiotensin II:* This is a key protein that helps regulate blood pressure and even controls blood flow to the fat cell itself. It's also part of the system that controls water (fluid) balance.

• *Tumor necrosis factor-alpha (TNF):* TNF is an important signaling molecule involved in varied functions related to fat burning, immune function, and cell death. People who suffer from low production of TNF are vulnerable to an array of diseases including Alzheimer's, cancer, major depression, and inflammatory bowel disease.

• *Insulin-like growth factor 1 (IGF-1):* IGF-1 is a hormone similar in molecular structure to insulin. It plays an important role in childhood growth and continues to have effects in adults, where it also promotes cellular growth and construction of certain proteins and tissues. (Widely marketed by antiaging clinics, IGF-1 has been popular among athletes for the same reasons that its better-known cousin, human growth hormone, has: It is believed to help make a human body faster and stronger by boosting muscle growth, reducing fat, and improving endurance.)

• *Interleukin-6:* Interleukin-6 belongs to the class of inflammatory cytokines. Cytokines are pivotal to our immune system, and are simply molecules that conduct certain signals to address illness or injury and spur the healing process. Rare deficiencies of a number of them have been well documented, all featuring autoimmune diseases or immune deficiencies.

• *"Good guy" prostaglandins:* These encompass a group of lipid compounds that are derived enzymatically from fatty acids. Among their strong physiological effects is regulation of the contraction and relaxation of smooth muscle tissue. They differ from regular hormones, though, in that they are not produced at a discrete site. Instead, they are made in many places throughout the human body, usually near where they are supposed to function. (A bit of trivia: The name *prostaglandin* derives from the prostate gland. When prostaglandin was first isolated from seminal fluid in 1935, it was falsely believed to come from the prostate. Prostaglandin activity is usually to blame when it comes to menstrual cramps in women, and why aspirin is often used to remedy the pain—in 1971, scientists discovered that aspirin-like drugs could prevent the production of prostaglandins.)

• *Nitric oxide:* Not to be confused with nitrous oxide (an anesthetic) or nitrogen dioxide (a greenhouse gas), nitric oxide is another important cellular-signaling molecule involved in many physiological and pathological processes. It's one of the few gaseous signaling molecules known to us. Among its many jobs in the body is conveying information between cells and increasing blood flow by dilating blood vessels. For this reason, it's sometimes given in supplement form to heart patients via nitroglycerin and amyl nitrite, which are converted to nitric oxide in the body. Nitric oxide also plays a role in the hair cycle, and, believe it or not, penile erections. Sildenafil, popularly known by the trade name Viagra (and similar Levitra and Cialis), stimulates erections primarily by enhancing signaling through the nitric oxide pathway in the penis.

- *Acylation stimulating protein (ASP):* This hormone is key to fat storage; we'll see shortly how it fulfills a major role in fat metabolism.

I could go on and on about the various molecules pumped out of fat cells, but the ones I've outlined give you a general sense of just how active fat tissue is in our body. Although some of these can indeed be associated with metabolic problems and certain conditions when they are produced in excess, they nonetheless comprise an essential tool kit for our health and longevity.

Fat metabolizes hormones. A multitude of life-sustaining hormones get processed in fat cells. In both men and women, testosterone, for example, is converted to estrogen in fat cells. In fact, most of the estrogen in men and postmenopausal women comes from the conversion of testosterone in fat cells. This helps explain why men who carry more body fat tend to take on some female characteristics like larger breasts (due to increased estrogen), while at the same time losing libido (due to decreased testosterone).

The metabolism of other hormones such as DHEA (dehydroepiandrosterone) and androstenedione also occurs in fat cells. These are important steroid hormones produced in the body that play into the production of sex hormones. (You may have heard about DHEA in supplement form. Because it's a precursor to estrogen and testosterone, DHEA is used for a variety of conditions, including osteoporosis, depression, and chronic fatigue, and to slow or reverse aging, improve thinking skills in older people, and curb the progression of

Alzheimer's and Parkinson's diseases. Athletes and other people use DHEA to increase muscle mass, strength, libido, and energy—though it's a banned substance by the National Collegiate Athletic Association. Truth be told, there isn't much evidence to support the claims of people who manufacture supplemental forms of DHEA. The best place to get this hormone is probably through one's own internal manufacturing plant—the one in which the fat cell is key.)

Last but not least, I should mention that the famous stress hormone cortisol is also metabolized in fat cells. Although cortisol is the victim of name-calling in health and weight loss circles, often labeled as a "bad stress hormone" that antagonizes weight loss, it's essential for energy metabolism and critical for survival. Cortisol is produced by your adrenal glands and stored in an inactive form in your subcutaneous fat. As your body fat increases, a certain enzyme increases in the fat cell to convert cortisol to its active form, which is then released from fat tissue into the blood.

Although we often hear that the heavier you are, the harder weight control will be, since fat cells pump out a lot of hormones that favor fat storage and retention (e.g., cortisol), let me again play devil's advocate and suggest that we may be jumping to conclusions. Bodily hormones become imbalanced long before fat cells help perpetuate this vicious cycle of fat retention. It's common today for many people to inadequately metabolize hormones, but not necessarily because they have too much fat. Numerous factors, many of which are unrelated to one's weight and body fat composition, can impact the metabolism of hormones, such as stress, poor nutrition, inactivity, and genetics. Excess hormones can then begin to circulate and get stored in the body, which over the long term can lead to more fat storage and an excess of certain hormones, particularly estrogen, in the body tissue. No doubt this imbalance can be worsened by an abundance of fat cells that further secrete certain hormones, but we can't disregard the influence of another huge factor acting on our endocrine's delicate system of checks and balances: obesogens.

Fat Cells Don't Make You Fat, At Least Not Initially

When the body's internal hormonal "wiring" gets a glitch in it somewhere, leading to hormonal imbalances, it's called an *endogenous imbalance*. This can be caused by any number of things, from a genetic trigger to chronic blood sugar imbalances. But when your body is exposed to exogenous, or external, hormones or hormonelike substances, then this, too, can throw your body off-kilter. Most everyone living on the planet today is frequently exposed to hormone-like substances that may be directly contributing to weight gain and stubborn fat. These substances, which are actually able to mimic hormones in the body, are now being called EDCs (*endocrine disruption chemicals*) or, more commonly, *obesogens*—a term coined by biologist and UC Irvine professor and researcher Dr. Bruce Blumberg. Examples of obesogens that we encounter daily include pesticides, herbicides, and fungicides found on conventionally grown produce and in municipal tap water. They are also present in conventionally raised meat and dairy products. Ironically, hormones in meat don't have much of a biological effect, but the antibiotics that farm animals are often injected with do, so these antibiotics are obesogens to us. We expose ourselves to obesogens in chemical form (e.g., bisphenol A, or BPA, perfluorooctanoic acid, phthalates, and parabens) when we use or consume from things made from plastic or vinyl, commercial beauty products, and many household goods.

Numerous well-respected studies have shown that endocrine disruption chemicals do just that: They damage normal function of metabolic hormones. They can, in fact, affect your body's signaling hormones that tell you when to eat and stop eating, and dictate how fast your metabolism runs and how your body fat acts—whether it favors fat retention or fat burning. One of the reasons we're concerned about what pregnant women are exposed to is that studies now reveal that obesogens are actually able to affect fetuses in utero. Obesogens that gain access to a developing fetus through the mother can target signaling proteins to tell the fetus to make more fat cells. And because we are born with a certain number of fat cells that stay with us for life, this obviously has lasting consequences. It not only increases the likelihood of body fat accumulation as a person ages, but it can also make weight management more difficult. The impact of obesogens on fetuses may in fact help explain the rise in childhood obesity. While we may cast blame on our children's propensity to eat poorly and prefer a more sedentary lifestyle, their inborn chemical makeup due to early exposure to these substances could be a more influential, primary factor.

Earlier, I mentioned how accumulated excess estrogen can have a negative impact in both men and women, a condition called *estrogen dominance*. Some EDCs—known as *xenoestrogens*—can act like estrogen in the body by binding to estrogen receptors, thereby encouraging estrogen dominance. And being estrogen dominant as a result of obesogens can sustain a cycle of fat retention and excess estrogen secretion. An overweight body can produce more estrogen by converting adrenal hormones into estrogen using a clever enzyme called *aromatase*. That estrogen triggers the formation of more fat cells, and the cycle perpetuates. This can be especially troublesome for middle-aged folks as hormone production shifts from their sex organs to other parts of the body, including the adrenal glands and fat tissue, a reality of aging we'll explore in the next chapter. And in a cruel twist of biological fate, the aging body struggling to maintain adequate hormone production will often hold on to fat, since it acts as a manufacturing plant for hormones.

It's important to note the role of blood glucose balance in all this, especially the critical link between insulin and body fat. As we'll see in the next chapter, many people are walking around with roaring blood sugars due to a diet high in processed carbs and sugar, which keep blood sugar levels up. And once insulin resistance establishes itself over time, this condition alone will promote fat storage. Making matters worse, insulin resistance can also cause irreversible damage to insulin receptor sites. So if you're insulin resistant for a long time, even the best diet may not help you to manage your weight easily.

{ CHAPTER 7 }

From Bagels and Broccoli to Body Fat: How Does It Get There (and How Does It Get Burned)?

Pick up any number of books promising to show you the quickest route to weight (i.e. fat) loss and you'll get an earful: Go low-carb; nix sugar and alcohol; try gluten-free; limit meat, dairy, and coffee; eat low-fat everything; attempt a fast; do a cleanse; and perhaps try this "Hollywood cookie diet." The sky's the limit when it comes to selling a weight loss strategy that promises quick results.

But most diets, whether they work for you, often fail to answer the ultimate question: How do calories lead to weight gain? Which calories are the "good guys" and which are the "bad guys"? What happens to the carbs, fat, and protein you eat when you down a goopy cheeseburger or raw kale salad?

First stop: your mouth. Saliva contains enzymes that break down starches in the food to simple carbohydrates, or sugars. From there it moves to the stomach along with any fat and water in the food. Once in the stomach, the mass of nutrients gets churned up and transformed with the help of certain substances like pepsin, an enzyme that digests protein, and hydrochloric acid. These chemicals further break down the food and turn it into chyme. The mixture then enters the duodenum, where the gallbladder secretes its bile to dissolve the fat in the chyme. Enzymes from the pancreas also rush in to further break down the carbs, fat, and protein. This process ultimately thins out the mixture into a fluid form so its nutrients can be more easily absorbed through the lining of the small intestine. Fats, carbs, and protein part ways.

The carbs have it easy: They rapidly hop onto the conveyor belt that is your bloodstream, and several different organs take the sugar they need as the blood passes by. Some gets stored in the liver as glycogen, and whatever is left is converted to fat and cached in fat cells. The fats also go into the bloodstream, but they are bound for the liver, which burns some of the fat and converts some to other substances (one is cholesterol), sending the rest to fat cells, where they wait until they are needed. The protein is broken down into building blocks known as peptides, which are further dismantled to become amino acids that can then move through the small intestine's lining and enter the bloodstream. From here, some of them build the body's protein stores while the excess is excreted. Superfluous protein stores are destined to become fat. Yes, even extra protein turns to fat. The idea that eating lots of protein will "put muscle on your bones" is flawed. Too much of anything will end up in fat cells. And I'll further remind you that excess body fat is not necessarily the result of eating too much fat, healthy or otherwise, but rather the combined result of a lack of exercise and overeating all nutrients, which includes proteins, carbohydrates, and fats. In other words, your body doesn't just take in dietary fat and use that to pad your hips, thighs, derriere, and abdomen. It makes its own fat from taking in excess *calories*.

Fat's Active Life: The Skinny on Fat Metabolism

So what happens to everything sent to our fat cells, from a biochemical standpoint? How does the body load up its fat cells and then tap them for energy when it's needed? Answering these questions through the story of fat metabolism will help us to understand an important lesson: Body fat is elegantly self-regulated to maintain our health, not prompt disease and dysfunction.

As you should see by now, even "bad" fats like LDL cholesterol and triglycerides serve their own important functions. First, some more basics, as this conversation, admittedly, gets complicated. As you know, lipids do not dissolve in water like other food items that enter the body, so they must first be broken down to be absorbed and distributed in the blood. Fats that you consume and fats that your liver can create from carbohydrates are transported by large molecules called *lipoproteins*. These lipoproteins are just that—a combination of fat and protein that can move in the bloodstream and communicate with cells. They are wrapped in a protein coat that contains specific receptors on the surface for binding to certain cells; hence, different lipoproteins are charged with different responsibilities depending on which cellular receptors they can work on. LDL receptors on a cell, for instance, will attach to LDL from the blood so the cells can extract cholesterol, while HDL molecules bind to receptors that allow them to take away "used" cholesterol—especially from arterial walls, where they get stuck in plaque—for recycling in the liver, such as when cells die.

Lipoproteins ferry fat around mostly in the form of triglycerides, which, again, are molecules made up of three fatty acids attached to a glycerol that acts as a kind of cohesive backbone to the fatty acids. More specifically, two particular kinds of lipoprotein carry most of the triglycerides: chylomicrons and VLDL (very low-density lipoprotein). Chylomicrons are created in the intestinal lining, where they collect and package digested fatty acids and cholesterol. From there they move on to the lymphatic system and eventually get dumped into the blood so cells can retrieve fat or cholesterol from them.

Although it's common to think of fat cells as these globular balls that can take in a lot of, well, fat, their doors don't accommodate large molecules. Triglycerides are too big to directly pass through a fat cell's outer membrane to get inside. What has to happen first is a little undressing: The individual fatty acids that comprise the triglyceride must be released so those smaller fatty acid molecules can then travel across the fat cell's membrane and land inside. The main enzyme responsible for this strip-down is lipoprotein lipase, or LPL. The glycerol molecule is left outside.

Now, here's where the story gets interesting again. As it turns out, fat cells don't just house fatty acids that they permit to enter through their cell walls. In addition, they build their own triglycerides with the help of glucose from the blood. So fat storage, to be clear, involves two crucial ingredients: LPL to liberate free fatty acids, and glucose. All of this activity is commandeered to a large extent by insulin, the body's chief metabolism hormone, whose primary job, as you know, is to shuttle glucose from the bloodstream into muscle, fat, and liver cells. If there's more insulin around (presumably because there's more glucose to manage), that means there will be more LPL and more glucose to transport across the fat cell's membrane—which ultimately translates to more fat storage.

Not all of LPL's mission is to load up fat cells. If it's attached to a muscle cell, for instance, it will drag fat into the muscle cell, where it can be burned for fuel. In other words, LPL hauls fat from the bloodstream into whatever cell happens to signal it. To complicate matters more, an enzyme called *hormone-sensitive lipase*, or HSL, awaits patiently *inside* the fat cell to conduct business going the other way. HSL does what LPL does but from inside the fat cell—it frees fatty acids from stored triglycerides so they can go into the blood and be used for energy. This is how your fat gets "burned."

Take note: In the absence of insulin, HSL responds in the absolute opposite fashion as LPL. This makes sense, because when there's less insulin around and more HSL activity, fat tends to be released so it can be used (burned) as a form of energy. But when insulin levels are high, LPL is in charge and fat tends to be stored. This exquisite cascade of events by which the two metabolic processes balance each other out provides the biochemical basis for the idea that overconsuming carbohydrates drives a lot of obesity. When we eat carbs, especially simple carbs that rapidly raise blood sugar, the body is suddenly swimming in glucose, and insulin levels rise, especially when this is not countered or neutralized with plenty of physical activity and exercise. This sets the perfect stage for fat storage and fat retention. As high insulin levels favor fat storage, activated LPL keeps those fat storages locked.

In addition to insulin, which has such a prominent role in our body, we owe a lot of our metabolic health to leptin, and looking at the entire hormonal system through the lens of leptin and its mantle helps us understand why we get fat, or, conversely, thin.

Leptin's Leadership

Leptin isn't your average hormone; it ultimately influences all other hormones and controls virtually all the functions of the hypothalamus, in the brain. Your hypothalamus is where your inner dinosaur lives; this ancient structure that predates humans is responsible for your body's rhythmic activities and a vast array of physiological functions from hunger to sex. Leptin is, at its most basic level, a primitive survival tool that holds the remote control to our metabolism and is uniquely tied to the coordination of our hormonal and behavioral responses to starvation. It also orchestrates our inflammatory response and can even control sympathetic versus parasympathetic arousal in the nervous system.

The more you can increase your body's sensitivity to this critical hormone through simple lifestyle choices, the healthier you will be. By "sensitivity," I mean how your body's receptors for leptin recognize and use it to carry out various operations. As I mentioned in the previous chapter, leptin helps control appetite. As such, it has a powerful effect on your emotions and behavior with regard to food and eating. The next time you put down your fork and pull away from the dinner table, you can bow to your leptin.

When fat cells start to fill up and expand, they secrete leptin, which acts as your brake at the table. Once the fat cells begin to shrink as their contents are burned for energy, the faucet is slowly turned off and less leptin gets released. Eventually you're able to feel hunger again and the cycle starts all over. This is but one example of the many different mechanisms the body has to expertly manage energy metabolism in the name of survival. People with naturally low levels of leptin are prone to overeating. A now seminal study published in 2004 showed how people with a 20 percent drop in leptin levels (due to sleep deprivation) experienced a 24 percent increase in hunger and appetite, driving them toward calorie-dense, high-carbohydrate foods, especially foods with a lot of sugar, starch, and salt. We've learned a lot about our hormones from sleep studies alone.

Now let's return to ASP, one of the hormones released by fat cells. ASP has two big effects. It increases LPL activity (resulting in delivery of fatty acids into the fat cells), and it increases the expression of those glucose transporters on fat cells so they ferry glucose inside the fat cell to result in fat storage. In short, ASP acts like insulin in fat storage but, unlike insulin, it's not produced by the pancreas. ASP is made by the fat cells themselves, prompting the production of triglycerides *inside the fat cells*. We think that this ASP secretion is triggered by the chylomicrons, those lipoproteins made in the intestinal lining that package up digested fatty acids and cholesterol. Remember, chylomicrons are essential for hauling dietary fats into the body, which is why eating a fatty meal leads to your body manufacturing more chylomicrons. These in turn cause the fat cells to make more ASP, triggering greater fat storage.

While we all know that we can eat enough of anything—fat, protein, or carbs—and become obese, we often forget that the body is more artful than we think. It has innumerable feedback mechanisms to try and keep itself balanced. We have evolved to be physiologically resilient over a wide range of potentially harmful conditions from our environment and nutritional lapses, and we are designed to store enough energy to survive in between meals and even periods of starvation without worrying about vital systems or our physical performance being compromised. All of the brilliant mechanisms that I've been describing shed light on just one part of an elaborate web of interactions that take place in the complex and endlessly mysterious human body. Notice how these reactions involve not just the metabolic system but also the endocrine, nervous, and digestive systems. I've barely mentioned the role of the brain in all this, but that's also part of the picture. Without the brain's ongoing intelligence activity, your body wouldn't be able to transmit hormonal signals at all, much less listen to the messages and perform certain functions, from the simple (overseeing the passage of food in your stomach) to the complex (coordinating intricate communications like telling you to say no to a second helping and push away from the table).

Houston, We Have a Problem

My whole objective in bringing up the fact the human body is much more complex than any of us could ever imagine is to drive home the point that obesity is rarely a condition that's "easy" to develop if you are blessed with a healthy body whose cylinders are all firing on cue, and you're not abusing it through lousy lifestyle choices. Usually, to become obese there is a glitch somewhere in this complex system that's big enough to perpetuate circumstances that favor unhealthy fat storage and, in the end, obesity. For example, a small handful of people in the world are born with a genetic defect whereby they don't produce enough leptin. This causes them to lack the "I'm full" signal. Luckily, these folks can be treated with leptin to reverse their rare condition. But sadly, what's much more common today is persistently high insulin levels coupled with a low sensitivity to the body's natural leptin secretions. These two forces together not only favor weight gain, but they also raze the body's metabolic health.

There's a reason why carbs bear the brunt of the blame when it comes to obesity. Carbohydrates, especially refined carbs and sugars, abound today, and many of them cause dramatic surges in insulin levels. This makes lots of glucose available for triglyceride storage. And even though insulin, in theory, should simultaneously work toward suppressing appetite, explosive levels will speedily shepherd energy nutrients into the cells from the blood, leading the brain to sense low levels of energy in the blood. This causes it to override built-in mechanisms designed to communicate fullness, such as the one led by leptin. The brain is, in a sense, duped by the effects of the violent insulin surges and may think starvation is on the horizon (the brain needs a certain level of blood sugar to be maintained in the blood just to ensure its own proper functioning). So what does it do? It activates signals that compel you to eat more, and you'll gravitate toward those carbs and perpetuate a vicious cycle.

Eating a balance of carbs with healthy fats and proteins does have its metabolic merits: It doesn't foment a perfect storm of insulin overdose that creates an imbalance in the body's hormonal system. Proper energy levels in the blood can be maintained and your preprogrammed system for controlling your appetite and managing hunger can operate normally. Although you could become obese by eating too many calories from fat, that would be difficult to achieve in reality because you'd have to force yourself to eat way past a fierce sensation of fullness, repeating that over and over again. Put another way, you'd be combating long-established systems in the body to prevent that fate.

I'd like to point out one more interesting fact about fat metabolism and potential spots where the system can be vulnerable to breakdown, which will help us further understand how body fat location can be important in terms of obesity's effects. It has to do with LPL, the enzyme responsible for cutting up fats outside cells so fatty acids can gain entry into those cells. The studies examining the role of LPL date back to the 1970s and have illuminated a powerful relationship between this hormone and another critical hormone, estrogen. We know now that estrogen interferes with the activity of LPL on fat cells. Put simply, if there's more estrogen around, less fat will accumulate in fat cells because the volume will be turned down on LPL. But once you take estrogen out of the picture, the fat cells will grow bigger as fat gets drawn into them. (The earliest experiments showing this effect were done on rats whose ovaries—the main organs of estrogen production—had been removed.) The researchers who first documented this relationship also noted how the lack of estrogen caused the animals to get much fatter than normal; they had an uncontrollable urge to eat of epic proportions (literally) because they were losing calories into their fat cells that were otherwise needed by the body elsewhere to keep things running.

Gary Taubes describes this unfortunate cascade perfectly in *Why We Get Fat*: "The more calories [the animal's] fat cells sequester, the more it must eat to compensate. The fat cells, in effect, are hogging calories, and there aren't enough to go around for other cells. Now a meal that would previously have satisfied the animal no longer does. And because the animal is getting fatter (and heavier), this increases its caloric requirements even further. So the animal is ravenous, and if it can't satisfy its newfound hunger, it has to settle for expending less energy."

One thing that Taubes spends a great deal of time asserting, and with which I agree, is that body fat is expertly regulated such that obesity cannot be explained entirely by behavioral vices rooted in the brain. It's sad to think that every conversation about obesity typically revolves around lack of restraint, gluttony, and sloth, not to mention that the main cause of progressive weight gains in our society over recent decades, as shown in several studies by my colleagues and me, has been the monumental reduction in calories burned from physical activity. Even though the obesity disorder is about abnormal growth of fat tissue, that narrative rarely includes a look at the hormones and enzymes that manage that growth to begin with. But now that we have so many more clues to understanding fat metabolism and phenomena like the obesity paradox, it's time we changed the narrative.

Humans are unlike any other species on the planet. We are the only ones who can actually get fat! A wild animal with plenty of physical activity living with an abundant food supply will never become obese. Yet it appears that we have evolved to be more sensitive to defects and malfunctions in the system, some of which can start relatively subtly and silently. And as this next chapter shows, for millions of people today, the initial misfire, or series of misfires, starts in the blood—not necessarily in the fat cell.

{ CHAPTER 8 }

Cause and Effect: The Real Killer Isn't Obesity

Just as I was about to dive into writing this chapter in August 2013, the following headlines hit:

> *Heavy Burden: Obesity May Be Even Deadlier Than Thought* (NBC News)
>
> *Obesity Kills More Americans Than Previously Thought: One in Five Americans, Black and White, Die from Obesity* (*ScienceDaily*)
>
> *Obesity Kills More Americans Than We Thought* (CNN)
>
> *No More Denial: Obesity Kills* (EverydayHealth.com)
>
> *Obesity's Death Toll May Be Higher Than Thought* (WebMD.com)

The news was based on a study recently published in the *American Journal of Public Health* suggesting that obesity accounts for about 18 percent of all deaths in the US between the ages of forty and eighty-five, which is three times previous estimates (existing literature places obesity-related deaths at only 5 percent of all US mortalities). It was controversial, to say the least. Not only did the study insinuate that the government has underplayed the dangers of obesity, but it brought to light just how difficult it can be to calculate the costs of being overweight. This wasn't the first time the government found itself in the center of the heated debate over how much obesity influences mortality.

In 2004 the Centers for Disease Control and Prevention retreated on its earlier declaration that obesity was something to take seriously, and that it was at or near the top of the list of leading causes of death, right behind tobacco. After estimating that 365,000 deaths a year could be related to obesity, the CDC second-guessed its analytical methods and cut that number down to 112,000. Obviously, the plot has thickened since the discovery that extra fat isn't always a death knell and may actually increase longevity, especially among those who already have a chronic condition that entails a higher mortality risk. It's a logical conundrum: If obesity is associated with heart disease, and heart disease causes premature death, then how do we reconcile the fact that some forms of obesity can help a person with heart disease live longer? Put another way, if you are six feet tall and weigh two hundred pounds, the BMI chart says you're overweight, but who can say for sure that you'll live longer if you drop ten or twenty pounds? Questions like these have agitated the obesity research community. They've also made it difficult to quantify "obesity-related deaths."

The impact of obesity has been prone to controversy ever since the CDC changed its estimation on obesity's overall death toll and new research began to emerge, bringing into question not only the ramifications of obesity, but the definition of obesity itself. Then, when Katherine Flegal's 2013 CDC paper hit (see chapter 5), finding fewer deaths among those deemed overweight than those in the normal range, health care professionals scratched their heads again. To some clinicians, such findings seemed shocking and implausible, since excess fat had been strongly and indisputably associated with an untold number of health hazards.

Let's go back to the *American Journal of Public Health*'s 2013 study for a moment, because it raised a few interesting points. Its lead author, Ryan Masters, PhD, posits that we overlook generational differences in the way the obesity epidemic affects us. For example, because younger individuals will be exposed longer to risk factors for obesity, they are at even greater risk of becoming overweight or obese. And as a result, they are also at risk for suffering from related health problems, and for a longer period of time. Kids growing up today are living in a totally different environment than even a generation ago. Obesity has become the norm, and so has a culture that feeds obesity's vicious cycle, characterized by its supersized meals and drinks, environmental obesogens, and ubiquitous access to caloric foods whose processed contents can aggravate stable metabolic processes (think blood sugar balance, blood pressure, and the like). Over the past five decades, we've progressively lowered the number of calories burned per day, which itself is a major cause of metabolic dysfunction and inevitable weight gain. People in their eighties today didn't know what obesity was when they were children because it was very rare. In contrast, children who will be in their eighties seventy years from now are growing up amid epidemic childhood obesity. The adverse effects of obesity will have the opportunity to encumber them over the span of all those decades.

To arrive at these new numbers on obesity's lethal toll, the researchers first broke the population down into generations, or "cohorts," and looked at the effect of obesity on deaths for different age groups. Using these cohorts, they studied the results from the National Health Interview Surveys from 1986 through 2004, which included the weight and height of those surveyed—numbers that were used to calculate body mass index and indicate whether someone was overweight or obese. Then they compared those findings with individual mortality records from the National Death Index, focusing on ages forty to eighty-five in order to exclude deaths caused by accidents, homicides, and congenital conditions (the leading causes of death for younger people). Within that frame, roughly one in five deaths among adults in the US was due to obesity (more specifically, they found that obesity was the cause of 20 percent of deaths among women and 15 percent of men). They also managed to determine that women seem to be more vulnerable than men to dying from obesity. Black women had the overall highest risk of dying from obesity or being overweight at 27 percent, followed by white women at 21 percent.

But we must be careful about taking these new estimations at face value. While younger people who are obese do indeed have a higher risk of dying early, the association between weight and mortality is surprisingly weak when we examine all the evidence and try to tease the details from the data. Why is the link between obesity and death risk so hard to figure out? In part, it's because obesity isn't a direct cause of death. And therein lies the biggest complicating factor: the difference between causation and correlation.

If you look up the top ten causes of death in this country, you won't find obesity on the list. Heart disease reigns at number one, killing more than half a million people a year, trailed closely by cancer (and by the time you read this, it's possible that cancer will have overtaken heart disease). Now, we can certainly say that obesity could have played a role, however big or small, in some of these deaths, not to mention many others on the list such as lower respiratory disease (no. 3), stroke (no. 4), Alzheimer's disease (no. 6), and diabetes (no. 7). But death certificates must list an immediate cause of death attributable to a single disease. So if you die of a heart attack and are sixty pounds overweight, we don't say you "died of obesity" or even "complications from obesity." We say you died of a heart attack. Period. (Death from obesity, per se, is exceptionally rare; it would require an extreme degree of overweight to be even plausible.) Casting obesity aside for a moment, let's face it: Many disparate things could have contributed to that heart attack. What if you smoked a pack of cigarettes a day for the greater part of your life? What if you loved steak and loathed exercise? What if your dad succumbed to a heart attack at forty-two and probably passed on that genetic vulnerability? These are just some of the more granular details that can quickly muddy the bigger picture we all see at face value.

We can continue this conversation in oh so many ways. As we've learned, it's well documented that people who are overweight or obese are far more likely than thinner people to have any of a number of diseases, but that being fatter does not necessarily mean being less fit or less healthy. We have fit but fat folks and the physically thin but metabolically obese. And we know that exercise and fitness can clearly protect you from death and disease regardless of weight. In Masters's study, he didn't account for smoking or other risk factors like alcohol consumption, much less protective factors, such as physical fitness. Nor did the study consider people's health insurance status, which contributes to mortality rates. The more access one has to good health care, the more likely one is going to be healthy and stay on top of health issues. What's more, many studies have found that the more educated people are, the more money they earn and the less likely they are to be obese. A third of American adults who make less than fifteen thousand dollars per year are obese, compared to about a quarter of those who bring in at least fifty thousand dollars per year. No doubt, this constitutes more confounding factors—perplexing circumstances that do more to blur the picture than point to any conclusive finding.

Obesity cannot be exonerated. But it can be reframed. To that end, I'd like to use the rest of this chapter to establish the following statement: Obesity doesn't directly cause any conditions known to kill. It is the middleman with many faces; it's nothing but a pathway variable that can exacerbate existing conditions and contribute to premature death by aggravating chronic disease—by adding more fuel to the proverbial fire already brewing in the body. If it killed in a straightforward, linear fashion, then we wouldn't have so many obese people who live long and who never get obesity-related disorders. And we wouldn't see so many thin people suffering from obesity-related conditions.

Cause and Effect: Causation vs. Correlation

It must be human nature to prefer things to be unambiguous, simple, and easy. The media in particular tends to be very good at keeping headlines as direct and pithy as possible (often at the expense of accuracy). To begin this conversation, I'll pick on one of my own studies, which came out shortly after the *American Journal of Public Health* study. The headlines generated for this one went something like: FOUR CUPS OF COFFEE A DAY MAY RAISE EARLY DEATH RISK IN YOUNGER ADULTS; DRINKING FOUR OR MORE CUPS OF JOE PER DAY CAN BE DEADLY FOR SOME; and even more to the point: COFFEE HABIT RAISES EARLY DEATH RISK. True? Blown out of proportion? Does coffee kill? I'll let you be the judge.

To say these headlines got people's attention is an understatement, and soon I found myself on a plane to New York so I could explain on broadcast television. Our decades-long study, which was published by the *Mayo Clinic Proceedings*, found that drinking twenty-eight cups of coffee or more per week is correlated with an increased risk of premature death by 21 percent. The risk was more than 50 percent higher in adults under fifty-five years old. (For the record, the average American adult drinks about three cups of coffee per day, which amounts to twenty-one cups weekly, not too far from this twenty-eight-cup threshold.) The debate about coffee's benefits and risks has swirled for years, with the current wisdom trending toward the positive side, as numerous reports reveal coffee's health benefits, including reduced risk for stroke, heart failure, Alzheimer's, some cancers, and even chance of suicide. In fact, Dr. James O'Keefe and I published a major review in 2013 in our top clinical cardiovascular journal, the *Journal of the American College of Cardiology* or JACC, describing the cardiac safety and several health benefits of drinking modest doses of coffee (e.g., 2 to 3 cups per day).

For the *Mayo Clinic Proceedings* research study, my colleagues at the University of South Carolina and I evaluated data collected between 1979 and 1998 on nearly forty-four thousand people between twenty and eighty-seven years old, most of whom were men. They all had answered health surveys about their medical history and lifestyle habits, including how much coffee they drank. Mining the data, which tracked deaths, we found that younger men who downed about twenty-eight cups weekly were 56 percent more likely to die prematurely from any cause than those who abstained from coffee. On the women's side, younger women who consumed that much coffee doubled their mortality risk compared to those who didn't.

But there were, as there so often are, confounding issues or additional variables to consider. For one, the people who drank larger amounts of coffee were also more likely to smoke and be less fit (as measured by treadmill tests). These circumstances alone could have added to their health challenges. (We can also reliably argue that people who drink this much caffeine often work in high-stress jobs and sleep poorly, both of which contribute to their health risks in a variety of ways.) Admittedly, the link we identified between coffee consumption and death risk in people older than fifty-five wasn't statistically significant, meaning the increased numbers could have been due to sheer chance. We also noted that the link didn't seem to increase one's cardiovascular death risk. And surprisingly, this remained true even though we know that excessive caffeine intake raises heart health risks because it increases a person's heart rate and blood pressure (which does become a greater liability as people age).

So does coffee kill? No. Can it be correlated with higher death risk? Maybe. But certainly not for everyone, and not unless we're talking about excessive amounts (although the media would have you believe otherwise at first to get you to tune in or read).

Dr. Leslie Cho, director of the Women's Cardiovascular Center, at the Cleveland Clinic, stated it perfectly for CBSNews.com when talking about the risks: "It's so much more than just coffee. The focus should be on moderation." And I'll add that the same could be said for obesity's risks. Its myriad perils involve so much more than just obesity.

If I were to ask you if junk food causes obesity in kids, could you answer a steadfast yes or no? By now you should know that we shouldn't jump to conclusions like that. If you thought about it, soon enough all the potential variables to reckon with would come to mind. We used to think that ulcers were directly caused by stress and spicy foods, as we noticed a strong correlation between the independent variables (i.e., stress and spicy foods) and the dependent variable (i.e., ulcers). But we know now that most ulcers are in fact caused by the bacterium *H. pylori* (a discovery that earned tenacious Australian scientists Barry Marshall and Robin Warren the Nobel Prize in 2005 for challenging prevailing dogmas and shedding new light on this particular bacterium's role in gastritis and ulceration of the stomach or duodenum, also called "peptic ulcer disease"). Although natural stomach acids and spicy foods can irritate an already damaged stomach and small intestine in people prone to ulcers, they don't directly cause the ulcers.

"Correlation does not imply causation" is a phrase used in science to emphasize that a correlation between two variables does not necessarily mean that one directly causes the other. In a widely cited example, numerous studies once showed that women who were taking hormone replacement therapy (HRT) had a lower incidence of coronary heart disease. This led doctors to suggest that HRT protected against heart disease. But randomized controlled trials—the gold standard in science—revealed something totally different: HRT caused a small but statistically significant *increase* in risk of heart disease. This prompted scientists to reconsider their thinking and take another look at the earlier data. That's when they found that the women in those initial epidemiological studies who were undertaking HRT were more likely to be from higher socioeconomic groups. In other words, they were more likely to practice better-than-average diet and exercise habits and to receive better medical care. So the use of HRT and decreased incidence of coronary heart disease were coincidental effects of a common cause. The superficial relationship could be attributed purely to the benefits of a higher socioeconomic status rather than cause and effect.

As with any logical fallacy, finding a flaw in an argument does not necessarily imply that the resulting conclusion is false. In the case of HRT, if the trials had revealed that hormone replacement therapy caused a decrease in coronary heart disease, but not to the extent suggested by the earlier studies, we could assume the causality to be true. Nonetheless, our logic behind the reasoning would still have been flawed.

One way to grasp the difference between causation and correlation is to turn to economics, which can be enormously useful in evaluating the quality of information in any situation.

In 2013, award-winning economist Emily Oster wrote a book called *Expecting Better: Why the Conventional Pregnancy Wisdom Is Wrong—and What You Really Need to Know*, in which she disproves many of the standard recommendations for pregnant women by looking at the data and showing how the real facts in the medical literature are often twisted by the time they reach the public and become more or less canons of law. In doing so, she explains the magnitude of distinguishing between causation and correlation by describing an example close to home: her husband's examination of the impact of television on children's test scores. Although conventional wisdom says television is often "bad" for young children (to wit: the American Academy of Pediatrics says there should be no television for children under two years of age, basing this recommendation on evidence provided by public health researchers who have shown time and again that children who watch a lot of TV before the age of two tend to perform worse in school), Oster's husband, Jesse Shapiro, also an economist, wanted to delve further into the facts behind the reasoning. He, like so many of us, was used to reading headlines in places like the *New York Times* that shouted out SPONGEBOB THREAT TO CHILDREN, RESEARCHERS ARGUE, but was skeptical of such aggressive statements. Oster employs an instructive example to recap her husband's research: Picture two families, each with a one-year-old, but one of the kids watches four hours of television a day while the other sees none. These families are probably dissimilar in ways that go beyond their children's television habits, for the kinds of parents who limit or forbid television tend to be older, more educated, and well-read, among other things. While we can say that television and test scores are *likely* correlated, that doesn't imply causation. A kid who watches television and who also has low test scores cannot blame just the boob tube.

Similarly, the idea that SpongeBob is solely responsible for making a child dumber is a causal claim. To prove that if you do X, Y will happen in this scenario, you'd have to force the kids who aren't allowed to watch television to actually watch SpongeBob and then show that they perform worse in school (while not changing anything else about their lives). But even if you could conduct such a challenging experiment, arriving at this conclusion would be exceedingly difficult just by comparing those who watch TV and those who don't.

To draw a causal conclusion, Shapiro and his coauthor on the study came up with a clever experiment based on the fact that when television was first getting popular in the 1940s and 1950s, it didn't land everywhere. There were some pockets of the country where television arrived first, so the authors identified children who lived in those lucky areas before they were two years old, and compared them to children who were otherwise similar but lived in places with no access to television until they were older than two. To keep other variables at bay, they made sure that the families of these children were all similar with the exception of the television experience.

What did they find? That poor SpongeBob and his counterparts were probably taking the fall for parents. The research in public health circles about the dangers of SpongeBob had been incorrect. The kinds of parents who let their kids watch a lot of television are probably to blame for a medley of problems. Correlation, yes. Causation, no.

The subject of obesity, like SpongeBob, suffers from a lot of misinformation. One or two weak studies can rapidly become conventional wisdom. And we can find ourselves making very broad, unsubstantiated claims that reflect false cause-and-effect relationships. Perhaps the best way to see how the conversation about obesity is often skewed by misconceptions about causation and correlation is to consider one of the most intriguing outliers to the obesity epidemic: people who develop diabetes—one of the most prominent conditions related to obesity—without becoming obese. It's the thin-man's diabetes, and more than anything it reveals that our most ruinous, pernicious disease today isn't obesity. It's chronic blood sugar chaos. The fact that not all obese people get diabetes and not all diabetics are obese means something else is going on. Indeed, appearances don't always reflect reality. And while we tend to cast blame on weight alone in discussing the chance of getting diabetes or dying from it, we forget the power of fitness in significantly reducing both of those risks. Moreover, blood sugar chaos isn't necessarily brought under control through diet and weight loss; physical fitness has a big say in the matter, if not the biggest say.

Blood Sugar Chaos Before Body Weight Chaos

It's long been established by many irrefutable scientific studies that a strong correlation exists between obesity and the onset of type 2 diabetes. We know that obesity is the major driver of type 2 diabetes among adults, and the reason type 2 diabetes (formerly "adult onset") occurs in children at all. It's also a documented fact that diabetes is a major contributor to cardiovascular disease, and that cardiovascular disease is the leading immediate cause of death among women and men alike in the United States. But acknowledging a connection between obesity and diabetes is one thing. What's less of a sure thing is the answer to the question of *how* obesity causes diabetes.

You'd think we'd know the answer to this seemingly "easy" question. But we don't. Although some new studies are pointing to a direct connection via a glitch in the immune system related to a certain protein, it's been an exceptional challenge in research circles to study and explain the exact mechanism by which obesity causes diabetes, because similar conditions in patients do not produce the same outcome. In other words, being obese is not enough in and of itself to cause diabetes. There's no exact mechanism by which obesity causes diabetes in everyone. One thing we do know, however, is that progressively higher levels of fitness are associated with progressive reductions in the prevalence of diabetes (this fact became the subject of a major paper published in the *Archives of Internal Medicine* in 2005, authored by Drs. Timothy Church, Steven Blair, and me).

Most people wrongly assume that diabetes (and, for that matter, high cholesterol) is a "fat person's" problem. This couldn't be further from the truth. As I've stated several times throughout this book already, plenty of people with killer-looking bodies are, in fact, facing the ultimate killer of all: diabetes. It's a narrative that doctors are increasingly hearing: A patient with the "perfect" body (for purposes of this example, let's say she's a 5'8" thirty-year-old woman who weighs an enviable 125 pounds) who comes into the office complaining of crippling, ongoing fatigue that cannot be explained by a passing cold or flu. Worries of a thyroid problem or cancer are quickly settled through tests. But the red flag shows up in the lab reports of her blood sugar levels, which are raging out of control. The patient is on the verge of diabetes. She is suffering from, as some call it, "skinny-fat syndrome"; or, as I noted earlier, TOFI: she's thin on the outside and fat on the inside.

At last count, nearly twenty-six million children and adults in the US—8.3 percent of the population—have diabetes, including seven million who have yet to be diagnosed. And here's the real shocker: An estimated seventy-nine million of us are walking around with prediabetes, just waiting for the day our body decides to downshift into a full-blown diabetic state. The CDC estimates that if current trends continue, one in three people will be diabetic by the year 2050. For decades, typical type 2 patients were close to what many of us still picture in our minds: heavy, inactive, and older, often receiving a diagnosis in middle age or beyond. But while such type 2 cases continue to climb, there has been a disturbing increase in a much younger, thinner set.

In the past decade, the number of people in their thirties who've been hospitalized for diabetes-related problems has doubled, with women 1.3 times more likely to be admitted than men. It's amazing to think that a serious condition that can often take half a lifetime to develop is suddenly a young person's problem. As I called out earlier, about 15 percent of people with type 2 diabetes aren't even overweight, a statistic that's aligned with the numbers of normal-weight people who suffer from serious metabolic abnormalities. Their "normal" bodies are hiding a dark secret. Although a much-talked-about common characteristic among people with type 2 diabetes is being overweight or obese, out of one hundred obese people, statistically only eighteen will develop type 2 diabetes. Obviously, something beyond excess weight is causing those eighteen people to develop diabetes and the other eighty-two to remain in the clear. And look no further than the cautionary tale of Jeff O'Connell to put this all into perspective. O'Connell wrote an eye-opening article in 2008 for *Men's Health* aptly titled "The Thin Man's Diabetes," with an even more appropriate subtitle: "America's fastest-growing disease has a sugar-coated secret: You don't need to be overweight for it to kill you."

For O'Connell, it started with a farewell visit to his dying father, who was withering away in a nursing home from advanced diabetes that had already claimed a leg. O'Connell's body, at 6'6", 220 pounds, and 12 percent body fat with a six-pack of chiseled abs atop a 32-inch waist, stood in stark contrast to his ailing father. In a twist of irony, just a week prior to seeing his dad, O'Connell had visited his own doctor after getting the call to review some blood work he'd had done several weeks earlier for a routine physical. The doctor had asked if diabetes ran in his family. O'Connell was well versed in the topic, for he'd been a writer at *Men's Health* for quite some time and had even cowritten a book on sports nutrition. But he was under the impression that diabetes was a disease destined to hit fat and old people, not people like him—he had a history of being thin and was even made fun of as a kid for being scrawny. O'Connell was in fact prediabetic, with a fasting blood glucose number around 116. (Under 100 milligrams per deciliter is considered good; anything above 126 is called diabetes. In 2004, the American Diabetes Association, or ADA, lowered the borderline glucose cut-point from 110 to 100. However, Drs. Mei Sui, Steven Blair, and I published a major paper in the *Mayo Clinic Proceedings* in 2011 showing that the stroke risk in forty-four thousand men really did not increase until fasting glucose values reached the 110 level, without significant increases between 100 and

110. Nevertheless, some of the adverse effects of blood sugar levels are noticed before the formal diabetes cut-point level of 126.)

Like most people who receive such a wake-up call, O'Connell immediately began thinking about how he hadn't sensed the danger, or even paid attention to the symptoms that had begun to bother him. He'd chalked up the chronic headaches, late-day exhaustion, and thirst to mere stress. After all, most everything can be blamed on stress these days, which can serve as a sneaky red herring. At the time, his doctor didn't offer much advice other than to switch from white rice to brown, and to come back in six months for another work-up.

The underpinnings of that first sign of disorder can start innocently enough, with just the consumption of too many carbohydrates or too little physical activity and exercise. Raging blood sugar is the initial hiccup in the body that can lead to a long list of ailments, from cardiovascular disease, liver disease, kidney failure, and stroke to amputations, blindness, nerve damage, and cancer. As O'Connell points out, there's not a substantial difference between the total amount of glucose in a typical person's bloodstream and that of someone crossing the border into diabetes territory. A healthy individual will have a little less than a teaspoon of sugar in the blood, while someone turning into a diabetic will have just about one-quarter teaspoon more. But such a seemingly scant additional amount of blood sugar is enough to make a huge difference in the body's physiology, sending your hormones on a roller-coaster ride. The ride begins with the consumption of carbohydrates that your body immediately goes to work breaking down so they can be absorbed into your bloodstream as glucose. As blood sugar levels increase in the bloodstream, the roller coaster begins its first incline, and the higher the blood sugar levels, the steeper the incline.

Although glucose offers a critical supply of energy for the body and especially for the brain, we know by now that excess glucose traveling around in your blood vessels is not an ideal situation. To adjust to a surge of incoming carbs, your pancreas secretes more insulin to usher all that glucose into your cells, where it belongs. This glucose leaving your bloodstream is the rapid decline of the roller coaster from the peak of the incline. As I've already described, the body runs into a problem when some of your cells lose their sensitivity to insulin and as a result deny access to glucose. This condition, insulin resistance, can go on for years undiagnosed. Eventually, this prediabetic state gets so bad and blood sugar levels remain so high that full-blown diabetes sets in.

Granted, the body is working hard to keep a semblance of balance throughout the roller-coaster ride. In the presence of excess sugar in the bloodstream, your body tells your pancreas to release more insulin to clear away that glucose. But as your pancreas pumps out higher and higher amounts of insulin, your blood sugar levels are driven sharply lower. This is when you feel like eating anything in sight, especially carbs, and may even feel shaky and light-headed. When you devour something to satisfy that primal need for food, your blood sugar soars again and stimulates another round of insulin pumping from the pancreas. The cycle repeats. But this isn't a mild roller-coaster ride; it's a violent one that goes up and down dramatically for a long time, sometimes years. And eventually the ride breaks down. The pancreas malfunctions from overuse and it stops releasing insulin. At this point, your blood sugar level remains dangerously high and you no longer have control over it. Of course, our factory-installed insulin system has worked fine for millennia. But today our pancreas hasn't had a chance to adapt to the onslaught of simple carbs from things like processed sugars, refined grains, and sugary beverages—the 140 pounds of sugar the average American polishes off annually. It also hasn't had ample time to adjust to our sedentary lifestyle.

Following O'Connell's wake-up call, he took matters into his own hands and went on a low-carb diet to see if he could rein in his blood sugar levels. He didn't want to go down the path of waiting for a bona fide diagnosis of diabetes and having to resort to drugs. And he brings up a good point: "Diabetes drugs are about as easy to ditch as crack."

Once you're on diabetes drugs, it's hard to go back. Most people end up stuck in a self-perpetuating cycle in which they need increasingly more drugs to control their blood sugar, especially if they aren't trying to cut the body's need for insulin by taking in fewer carbohydrates and boosting their blood sugar burn rate through exercise.

O'Connell eliminated foods such as pizza, cereal, bread, pasta, rice, potatoes, and desserts from his diet and instead turned to fresh fruits and vegetables, nuts, eggs, and meat. He also upped the ante by kicking his exercise regimen into high gear, for he'd read a study indicating that insulin resistance in rats decreased more from exercise than from taking metformin, the leading diabetes drug (which is true).

By his next doctor's appointment, things were looking much better. O'Connell's fasting blood sugar and blood fats, which tend to rise with blood sugar, had dropped to near-normal levels. The doctor was especially impressed by the transformation of his results on the hemoglobin A1c test, a three-month running average of blood sugar levels. Ideal levels for a nondiabetic are between 4 and 6 percent, and after months of following a low-carb diet and exercising more, O'Connell's results landed perfectly in the middle, at 5 percent.

But all was not perfect—far from it. Looking for a second opinion, O'Connell had sent his test results to another doctor, who noticed a major inconsistency. While O'Connell's A1c test results were normal, his fasting-glucose score was still too high. According to this new doctor, O'Connell's blood sugar was bottoming out and he was hypoglycemic much of the time (hypoglycemia is defined as less than 70 mg/dl; normal blood sugar is between 70 and 100 mg/dl). Put another way, O'Connell's blood sugar was surging upward and crashing downward in repeat cycles, and his perfect A1c score was averaging the extreme highs and lows.

This led O'Connell to take the next step in decoding his metabolic issues, which included a special kind of metabolic stress test, in which the doctor monitors and records the behavior of one's metabolism upon the consumption of sugar over time. And indeed, this three-hour test indicated that O'Connell had what's called *reactive hypoglycemia*, which, in his own words, "may be diabetes's most brilliant disguise of all." In someone like O'Connell, when he eats carbs, his blood sugar spikes to a prediabetic state, but because his insulin does a poor job of escorting sugar into cells, his pancreas ends up producing ten times more insulin than it should. This burst of insulin is more or less like an atomic bomb in his system, driving his blood sugar down over the next several hours until it hits rock bottom, about twenty points below where it started. (And this is when O'Connell is aching for a nap.) O'Connell sums up the crux of the misadventure:

"No wonder people are bonking at their desks all afternoon. Your brain produces no energy itself, yet it sucks up 25 percent of the glucose circulating throughout your body while you're up, and about 60 percent at rest. During hypoglycemia, gray matter is literally starving. . . . You become shaky, anxious, dizzy, sweaty, tired, and unable to concentrate. Your body does whatever's necessary to protect your brain, and that includes breaking down muscle tissue so that it can be converted to glucose. Which begins to reveal why someone built like my father or me could be fast-tracking his way to type-2 diabetes. Because our insulin resistance results in frequent periods of low blood sugar, our bodies spend a good chunk of the day eating our own muscle."

And that's exactly why some people can look so good and stay so thin yet hide such a deep, dark secret coursing through their bloodstreams. He writes: "In fact, insulin resistance is typically thought to cause weight gain, and vice versa. All of which makes the 'thin man's diabetes' that much more perplexing." And he's right.

For O'Connell, his cure meant avoiding not just the dangerous ups of his blood glucose level, but the equally menacing lows. So in addition to sticking with his relatively low-carb diet, he keeps his blood sugar levels on an even keel by eating frequently, and long before strong sensations of hunger commence. And lo and behold, at his next doctor's visit, everything looked normal. He writes: "Type-2 diabetes still lies waiting for me. It just needs me to drop my guard and eat junk food, put [too much] sugar in my coffee, skip meals, fall out of shape, and forget for even a brief stretch that this metabolic fire needs only its oxygen to roar again."

Taking Control of Your Master Switches

Without question, O'Connell's plight is an anomaly in the sea of stories about diabetes, most of which revolve around weight issues. But his experience brings up a lot of relevant points. Even though he didn't gain the kind of weight most people do when they bombard their bodies with too many carbs that in turn lay siege to the body's insulin-pumping system, his story goes to show just how destructive blood sugar imbalances over the long term can be—especially when combined with low levels of physical activity. Obesity didn't cause his diabetes; the anarchy in his blood sugar balance did. And the same can be said for the people who do in fact end up on the weighty side with type 2 diabetes. The point of origin isn't necessarily in the fat cells; it's in the bloodstream and much of the problem may be due to physical inactivity.

There's also no question that as the body's blood sugar metabolism begins to run amok and fat cells start to get loaded up, a classic vicious cycle commences. Once the body is locked into a position in which any gram of incoming carbs is poorly managed and every ounce of excess fat pulses with more fat-stuffing hormones that stifle weight loss, it's difficult to turn the train around. This is why it can be so hard to look at an overweight or obese person suffering from diabetes and not instantly blame the extra padding alone. It's just so visual that we forget what could have been going on "silently" long before the chunky rolls and love handles started to galvanize, grow bigger, and express their wrath.

Once fat cells are victimized over time by taking up too much slack from uncontrolled blood sugar, they, too, can wreak havoc on insulin regulation, via a road going the other direction—from the fat cell to insulin's command posts. We are now beginning to speculate that the risk for insulin resistance and diabetes also could be related to the number of fat cells a person has. According to researchers at Monash University, in Melbourne, Australia, this is because fat cells release a protein called *pigment epithelium-derived factor* (PEDF), which can trigger insulin resistance and pave the path to type 2 diabetes. And it follows that the number of fat cells that a person has is directly proportional to the amount of this protein that is released.

Other researchers studying how obesity itself causes diabetes have offered differing answers, many of which have something to do with substances released by fat cells. Research from the Joslin Diabetes Center, in Boston, for instance, has revealed that fat cells also produce a hormone called *resistin*, which essentially prevents cells from correctly responding to insulin (the cells "resist" insulin). This leads to the development of abnormal blood glucose levels, appetite, and fat storage. And yet another study, from the Harvard School of Public Health, has illuminated a process by which the part of the cell membrane that processes fats and proteins—the endoplasmic reticulum—is stressed by obesity. The endoplasmic reticulum can trump insulin signals by alerting the body to ignore insulin until it finishes its job of processing fats and proteins.

Long story short: A lot is going on in a body whose self-regulating metabolic processes have been dethroned. And that dethronement often entails a two-way street: We have the blood sugar maelstrom moving one direction and the pandemonium coming from fat cells in the other direction to make for one subversive highway.

This discussion brings up another question: How much of one's risk for diabetes is genetic? How much does your DNA dictate whether you'll become obese and suffer the metabolic consequences? In truth, probably a lot of your risk rests in your DNA. After all, your genetic code likely determines how many fat cells you're capable of producing, where you'll accumulate your fat, and how your body responds to your lifestyle. Each of the "master switches" in our bodies that dictate when and if we get *xyz* is likely under control by our personal genetics. But remember, although you didn't choose your parents, you *can* choose to fight back at your genetic predispositions via the powerful forces of nutrition and, particularly, physical fitness—both of which can indeed modify how your genes are expressed. And you can markedly decrease the chances that you'll develop metabolic problems and diabetes. As part III explains, it's all about striking the balance.

{ PART III }

Striking the Balance

{ CHAPTER 9 }

Walking the Fine Line: When Sweating Breaks Your Heart

The idea that too much of anything can kill extends to pretty much everything in life. Even water, which makes up about 66 percent of the human body, can turn toxic when consumed to such excess that it changes the chemistry of the blood. So as counterintuitive as it may sound, the same can be said for exercise (and as irony would have it, water intoxication happens to be more common among athletes in their attempts to rehydrate after an event: A 2005 study in the *New England Journal of Medicine* found that close to one-sixth of marathon runners develop some degree of water intoxication).

I intended this chapter to be short for fear of arousing worries that overexercising should be on most people's minds. It shouldn't; a much bigger problem for our society as a whole is too little exercise rather than too much. One of my concerns about playing with this debate is that I don't want to confuse those of you who may mistakenly hear me say that exercise is bad. Obviously, people who exercise are much better off than those who do not, and my colleagues and I have spent a lot of energy both academically in our careers and individually with our patients promoting the benefits of regular exercise and efforts to improve cardiorespiratory fitness. Physical activity is important—that goes without saying. As little as fifteen minutes per day confers substantial health benefits; forty to forty-five minutes five to six days a week is ideal, slashing your risk almost in half for premature death, diabetes, Alzheimer's disease, depression, and heart attack. People who regularly engage in physical exercise have notably lower rates of disability, and their average life expectancy is about seven years longer than that of folks who don't ever break a sweat. We have to make certain that this message does not become diluted by a discussion about when exercise becomes counterproductive. Nonetheless, I am compelled to disclose what I know about the limitations of exercise, if for no other reason than to emphasize the importance of moderation in everything. So when does fitness become a liability?

To answer that, we can turn to a longitudinal study published in *The Lancet* in 2011 that shows the law of diminishing returns when it comes to physical activity. Based on more than four hundred thousand people in Taiwan, the researchers clearly confirmed that physical activity is associated with reduced mortality: The more you exercise, the longer you're going to live. But there's a significant difference between people who exercise "vigorously" and those who engage in moderate activity. The more strenuous the exercise, the quicker one reaches the zone where the benefits of exercise start to diminish. You gain the most benefits from vigorous activity in the first forty to fifty minutes, after which additional time spent exercising doesn't necessarily offer more benefits and might actually cause harm. Conversely, a moderate exerciser can enjoy the rewards of exercise for a much longer period.

But it would be irresponsible to end the conversation here, because this latest news has stirred a lot of heated debate. So let's address the bigger question: Have we gone to extremes when we pursue sports like marathons and triathlons and engage in vigorous physical exercise for more than an hour at a time? The original marathoner—Phidippides, a forty-year-old herald messenger (professional running courier)—died delivering news of the Greeks' victory at Marathon (hence the name of the sport). Centuries later, fifty-eight-year-old ultramarathoner Micah True, the hero of the book *Born to Run*, collapsed in March 2012 while on a twelve-mile training run in New Mexico. Did a fateful glitch in his DNA switch his heart off, or did decades of such exertion lead him to develop so-called *Phidippides cardiomyopathy*—the umbrella term for multiple cardiac abnormalities that together lead to fatality?

Although True's passing went mostly unremarked upon in the media, studies conducted by me and my colleagues that emerged around the same time most certainly did not. The headlines hit all the major presses around the world, from the U.K.'s *Telegraph* (TOO MANY MARATHONS CAN KILL, WARN DOCTORS) to the *Wall Street Journal* (ONE RUNNING SHOE IN THE GRAVE: NEW STUDIES ON OLDER ENDURANCE ATHLETES SUGGEST THE FITTEST REAP FEW HEALTH BENEFITS). We now have a series of landmark studies showing that the benefits of running may come to a screeching halt later in life. I have been lucky enough to participate in many of these investigations as a lead researcher, so I have a unique take on their interpretation. I also had an insider view on all the fiery banter that took place following the publication of one hotly contested *Mayo Clinic Proceedings* paper, which I coauthored with one of my best friends and colleagues, Dr. James O'Keefe. Dr. O'Keefe hails from Saint Luke's Mid America Heart Institute, in Kansas City, Missouri, where he is revered for his groundbreaking studies in the field of cardiovascular medicine, diet, and exercise. (In addition to his cardiology work, he's a former elite athlete. Each year from 1999 to 2004, he won the largest sprint-distance triathlon in Kansas City.)

The gist of our findings, plus those confirmed by others, is that the negative effects of prolonged intense exercise and high-endurance activity include physical, structural, and biochemical changes to the heart and, in turn, the circulatory system that can increase one's risk for heart trouble. It's well-known, for instance, that athletes' hearts have to adapt to their rigorous exercise routines to meet their bodies' increased circulatory demands. This often entails an enlargement (overstretching) of the heart's chambers, thickening of its walls, and changes to electrical signaling. In the general population, such changes are associated with a higher risk of heart problems, but they are deemed "typical" in highly trained athletes and are not always threatening. The main issue is that these heart abnormalities, which pile up over a number of years, can predispose an individual to irregular heartbeats, which take their toll with increased age.

There's virtually no disagreement in the cardiology community that endurance athletes greatly increase their risk of atrial fibrillation, which is estimated to be the cause of one-third of all strokes. Older athletes (e.g., those fifty and above) who've been exercising long and hard for the greater part of their life may face elevated risks as they continue to age and fall prey to accumulated damage done in combination with the realities of aging. In fact, long-term excessive exercise may even accelerate aging in the heart; we've documented such symptoms, including increased calcification in the arteries and stiff arterial walls, in endurance athletes. What's more, changes at the biochemical level can further lower the threshold on one's overall risk factors for heart problems. With extreme exercise events, many athletes' hearts release substances often correlated with heart attacks and heart failure. Troponin, creatine phosphokinase (CPK), and brain natriuretic peptide (BNP) are three such chemicals. A dilation of the heart's chambers and weakening of cardiac contraction are frequently seen in extreme exercisers and heart patients alike.

The heart pumps about five liters per minute when we're sitting, and when we're running it catapults to twenty-five to thirty liters. The heart wasn't meant to do that for hours, day in and day out (not even when we were cavemen and -women scrambling to find food). Such demand on the heart leads to overstretching and tearing of muscle fibers. I don't think it takes a scientist to draw a few reasonable conclusions from those facts alone. In the lab, we can see the resulting thickening of the heart's walls and changes to its functionality.

We've in fact documented these subtle signs of heart damage in marathoners tested immediately postrace. As much as 30 percent of marathon finishers, for instance, have elevated troponin levels, irrefutable evidence of heart damage that cardiologists like me use (often in an emergency room) to determine if someone's having a heart attack. To be clear, the levels of troponin found in marathon finishers aren't anywhere near as high as those associated with heart attacks. Nevertheless, over time the damage adds up. Although these abnormalities seem to resolve over the next several days following an extreme event, there is concern that some individuals develop scarring of the heart that can lead to serious rhythm disturbances and even death, especially for aging athletes who don't give their hearts a chance to heal. MRIs done on longtime marathoners tell us that at least 13 percent and as many as 50 percent have abundant scarring on their hearts.

It's one thing to train for a single marathon you want to check off your bucket list, but it's quite another to make a habit of participating in serial marathons for years on end. While your body wants to exercise, it also wants to recover. A "lower-dose" runner is one who runs fewer miles, less often, and at slower speeds. Imagine a light jog around the park: It's enough to break a sweat, but not so exhausting that the body collapses. These lower-dose runners have been shown to have a significantly lower mortality rate than extreme runners (and nonrunners for that matter).

Perhaps the best proof of this delicate balance between true fitness and a health hazard comes from two large studies that have provided tremendous data on runners and life expectancy. One of those studies, which allowed me the great fortune of collaborating again with Dr. Steven Blair (I've mentioned him numerous times; he is often considered the "father of cardiorespiratory fitness" worldwide), has uncovered some decisive evidence. In our Aerobics Center Longitudinal Study, presented in mid-2012 by Dr. Duck-chul Lee at the American College of Sports Medicine, more than 50,000 runners and non-runners were followed for an average of almost 15 years. The runners in the group had a considerably lower death rate than non-runners. In particular, the lowest all-cause mortality rate was found among the runners averaging about 15 miles a week, and slightly higher mortality was found among those running well over 20 miles weekly (the higher risk was still lower than the mortality of non-exercisers). But among the running group, Dr. Lee reported that those who ran 25 or more miles a week seemed to partly lose that mortality advantage—they ended up with nearly the same risk as the lounge lizards in the study. Additionally, those who ran faster than 8 miles per hour or ran more frequently (6 or 7 days weekly) appeared to have fewer benefits than those who ran slower or less often (2 to 5 days weekly). Although a detailed dissection of this data raises the question of whether the high running doses (and 25 miles per week would not be considered a super-high dose among regular runners) really cause any toxicity, there is no question that maximal benefits didn't require the high doses but rather occurred at quite low running doses. Another colleague, Dr. Peter Schnohr, is the lead author of a major Copenhagen jogging study published in 2013 in the *American Journal of Epidemiology*, which showed findings somewhat similar to those in our paper from the ACLS: Runners lived longer than did non-runners, but the low-distance runners appeared to fare a bit better than the very high-distance runners.

Granted, these studies did have their own flaws that could have skewed the results. We can't rule out the possibility that some of the people doing the most running could have had preexisting cardiac issues that were worsened by the running or had very high-stress lifestyles, which the study didn't take into account. And certainly not every overly ambitious veteran athlete who's been racing for decades dies of heart disease. When we document heart damage in exercise enthusiasts, it's impossible to figure out how much of that damage is directly caused by the effects of pushing too hard in exercise and how much is due to an underlying genetic component. Moreover, in any large group of runners, there aren't many fast-paced people who run long distances to make any reliable conclusions that are statistically significant. But at the same time, we must respect the heart for what it is: a muscle. And like any other muscle in the body, it can sustain injury over the long term from overuse. Dosage is just as relevant to exercise as it is to any other medical treatment. Words of caution have even come from none other than Kenneth Cooper, the Dallas physician widely credited with launching the aerobics movement nearly half a century ago. He agrees that if you're exercising to extreme lengths, you're doing it for some reason other than health, and you could be rendering your body more susceptible to disease and dysfunction, including the most feared of them all: cancer.

I think the bottom line is evident, as I've concluded with my colleagues in my studies: No amount of light to moderate exercise is harmful. A routine of moderate physical activity will add life to your years, as well as years to your life. Clearly, everyone has a different definition for what determines "light," "moderate," and "extreme" exercise. And my guess is the vast majority of people reading this book don't come close to the level of extremity studied in these reports that reveal exercise's dark side. Even a relatively brisk walk for ninety minutes won't qualify as "extreme endurance exercise."

As I like to say in my talks, if we had a pill that does everything that exercise does, I'd be out of a lot of business as a cardiologist. But like any potent drug, an ideal dose exists that's just enough to get the full benefit but not enough to inflict harm or cause any dangerous side effects. And according to the most up-to-date data, I can confidently say that the ideal dose of daily vigorous exercise is about thirty to sixty minutes. You can gain the majority of health benefits just by walking briskly for twenty to thirty minutes daily. If you exercise strenuously for more than sixty minutes a day, you could potentially start to erase or even reverse some of those health benefits gained from lighter amounts of activity. Knowing the difference between average and excessive is easier than you think if you just listen closely to your body, especially as you approach middle age. Pay attention to signs of overdoing it, such as unusual heart palpitations, irregular heartbeat, chronic fatigue, and needing a long time to recover from exercise. Any of these symptoms should be a signal to you to make the appropriate adjustments to your routine. It shouldn't take a formal health chart or doctor to tell you when enough is enough.

I don't pretend I'm twenty anymore when I hit the pavement for a run (my marathon personal record was 3:10, a time I won't see again, though I enjoy five to ten races yearly). I used to average about

50 miles a week, but now my weekly mileage is closer to 35 or 40. On busy days, I might run 4 miles when I would have run 6 previously, and on leisurely weekends I'll run 9 or 10 miles, knowing full well that I'm getting most of the health benefits in the first 40 minutes or so. I continue to run for reasons other than health and stress management. In addition to being something that I truly enjoy doing, it also gives me time to think over my lectures and plan. (Dr. O'Keefe, also a runner, has similarly cut back on his mileage, slashing his running to less than 20 miles a week, keeping his speed at 8 minutes a mile or slower.)

Without a doubt, a sedentary lifestyle will most likely cause disability and disease, as well as a shorter life span (barring superhuman genes). This has been true for as long as humans have roamed the earth, for we were never designed to be an idle, stationary species. And we haven't evolved to sustain long, tough episodes of athleticism on a routine basis. Although we like to think our ancient ancestors ran all day long to survive, ethnographic research indicates that they were a tad less ambitious, walking a mere four to ten miles a day. Walking is arguably superior to running for a variety of reasons, and you can't, by any means, overdo walking like you can with running. (Dr. O'Keefe and I advise our patients that they can walk or garden hours a day without worrying about abusing or injuring their cardiovascular system.)

So while it's true that exercise provides numerous health benefits, the common belief that "more is better" is clearly not true. Modestly moderate—not even close to extreme—physical activity is ideal. And as I stated earlier, which begs to be repeated: The kind of exercises we need to do to win marathons and condition our bodies to be those of elite athletes are very different from those that preserve our health and support overall longevity. So, if your goal in life is to compete in the marathon or triathlon of the next Olympics, such an endeavor will certainly require lots of intense exercise for prolonged periods. Training for and participating in a single endurance event won't kill you, but you might not want to do multiple events a year over the course of several years. And as Dr. O'Keefe and I recently wrote in the journal *Heart*, if your goal is to be alive and well while watching an Olympic event decades from now, then physical activity at lower intensities and for shorter periods of time is more ideal.

{ CHAPTER 10 }

Your Personal Prescription: 10 Principles for Minding the U-Curve

Hippocrates, the father of medicine, taught, "The right amount of nourishment and exercise, not too much, not too little, is the safest way to Health." If you listen to your body, this is just common sense. Yet our culture is increasingly one of extremes. In my lifetime I've watched obesity rates triple throughout much of the Western world and I've simultaneously seen the number of people completing marathons rise twentyfold. People who live on either extreme end will always find ways to excuse themselves from living a truly healthy life in the safe and comfortable middle. This is the sweet spot where one engages in moderate regular exercise to maintain fitness and sustains a body weight that can support metabolic health.

Now that I've given you a wealth of new knowledge, some questions remain:

- In light of the obesity paradox, should doctors actually be advising overweight and obese patients to lose weight, or not?
- What about lean, "normal-weight" patients with cardiovascu-

lar disease: Should doctors be trying to promote weight gain or purposely trying to add fat?
- Does fitness totally and always negate the effects of fatness?
- Does belly fat matter if you are also fit?

These questions reflect the perplexing array of issues that arise around the obesity paradox. Although we all like our facts and directives about how to live a healthy life to be very clear-cut and straightforward, the real world is anything but. This chapter will squarely put these questions to rest while also achieving a big additional goal: telling you how to put all the information you've learned into action. Clearly, everyone will have come to this book with a different set of personal circumstances. Some will be tiptoeing close to a dreadful diagnosis like diabetes or heart disease regardless of their weight, while others might already be managing a chronic condition or have survived a heart attack or heart failure, or who have other heart conditions, such as hypertension or atrial fibrillation. And still, there will be those who epitomize the standard portrait of health but who would do well to heed this book's message to optimize prevention. You are curious about how to balance the competing needs for certain fitness and fatness levels while also potentially benefiting from weight loss, and how to safely go about doing that.

All of these recommendations are relevant for people with and without diagnosed conditions. At the end of the chapter I will specifically address those with established heart disease, since this is my area of expertise and it remains our number one killer.

The Obesity Paradox's Main Lessons

If we put all the obesity paradox evidence we've collected so far and laid it out on a table, you'd soon find yourself looking at a gigantic, intricate jigsaw puzzle (with a few missing pieces) that doesn't really have a single solution. You can assemble various pieces, but at the same time, you will find alternative solutions using different shapes that also seem to fit here and there. It's as if the puzzle is constantly shifting and changing its shape as you struggle to reach a finite endpoint. No sooner do you try to talk yourself through it than you wind up tongue-tied.

This is the obesity paradox. But despite its complexities—especially since each of us carries a unique physiology and set of risk factors both inherited and not—we can nonetheless extrapolate a few principles that everyone would do well to follow. The first ironclad rule is the one I've already underlined numerous times: Being at either end of the spectrum, either morbidly obese (BMI 40 and above) or painfully thin (BMI less than 18.5) spells trouble. It's a classic U-curve: You don't want to be out on the edges.

The benefits of fitness cannot completely outweigh the cons associated with these two severe body types; it's extremely difficult to be deemed "fit" if you're underweight or morbidly obese by BMI standards (rare exceptions include some gymnasts and dancers and, on the other end of the spectrum, NFL linebackers). And even though we know that the body mass index doesn't always tell an accurate story (i.e., to say in a blanket statement that obesity is bad and thin is good is fraught with errors and oversights), we do know that the law of diminishing returns rings loud and clear the closer one gets to these two extremes. Anyone who falls under one of these two categories would do well to focus on bringing their weight closer to the middle of the curve. Individuals with a BMI under 18.5 can improve their health immensely just by increasing their weight to move into the normal-weight category; similarly, the morbidly obese (which now comprises nearly 3 percent of the population) can benefit greatly just by losing enough weight to shift down into the moderately obese category (BMI 35 to 39). Unfortunately, the distribution of BMI in the US has shifted in a skewed fashion such that the proportion of the population with morbid obesity has increased by a greater extent than that of overweight and mild obesity.

We also can agree that even the term *obesity* begs to be redefined; it's not fair to lump someone with a BMI of 30 ("obese") who is metabolically healthy and from a cardiovascular fit standpoint in the same category as a person with a BMI of 39 who can barely climb a staircase without getting winded. By the same token, a fit person with a BMI of 28 ("overweight") could easily outlive a thinner individual (let's say BMI of 22, "normal weight") who is not fit, even though the prevailing wisdom is that anyone with a BMI of 25 or greater bears dramatically greater risk factors for things like cardiovascular disease, type 2 diabetes, and certain cancers. So while it's right to associate obesity with diabetes and other health problems, the revered BMI charts don't give us complete information, and the so-called "healthy" weight range may differ from person to person. In a perfect world, BMI would be able to incorporate other variables that define health, such as fitness levels, genetics, and biomarkers of metabolic health such as inflammation, blood glucose, and cardiovascular risk factors.

Remember, weight and metabolic health are imperfectly correlated, making the story of health within the framework of obesity all the more puzzling. While nearly a quarter of obese adults are metabolically unhealthy, which is to be expected, almost one-quarter of "normal-weight" people are also metabolically unhealthy. This encompasses a surprisingly large group of people that most physicians and public health authorities forget to think and worry about. In addition, research shows that more than half of "overweight" and almost one-third of mildly and moderately "obese" people are metabolically healthy, which amounts to quite a lot of individuals—fifty-six million Americans. These are the people who are chubby or "full figured" but not in danger of dying early, especially if they are in physical shape, yet are relentlessly censured by society—the media, employers, insurers, public health authorities, doctors—to lose weight (or else!). They are the target of the multibillion-dollar diet industry, not to mention all the other tools, gadgets, programs, surgeries, and supplements sold in the name of weight loss. Remember, too, that the science now shows that people in these BMI categories (25 to 34) may be labeled as overweight or obese by modern medical standards, but they may enjoy significantly lower mortality rates (particularly the overweight range 25 to 30) than do their normal-weight counterparts.

Does this mean you should be trying to gain weight and get into this club if you've got a BMI of 24? If you already have a diagnosed chronic condition often associated with obesity, and you're not in the severely overweight category, does the science merit a recommendation whereby you should avoid losing weight and actually *gain* body fat? Not in the least! Please do not misunderstand what I am saying or insinuate that I am promoting obesity. Nevertheless, I am saying that you'd do yourself a better favor by maintaining your weight and increasing your fitness, both of which will help you to avoid obesity-related conditions. But for those who are already part of this overweight/obese club—and it's a big one today in America—the message is that you can be as healthy as or even healthier than your thinner friends. And the fitter you are, the healthier you'll be. Even if you suffer from a chronic condition usually associated with extra fat, your level of fitness will protect you to some degree despite your condition. Which is why, when aging brings on stubborn extra weight, the prognosis remains good so long as you maintain fitness. In fact, some studies, which I covered earlier, show that if you maintain or improve fitness, changes in weight don't even matter, as my friends Drs. Steven Blair, D. C. Lee, Mei Sui, and Timothy Church have demonstrated many times over the years. When people start to develop exercise habits that improve their cardio fitness and muscle strength, everything else improves (i.e., body composition, overall health profile, risk for disease and death).

To the layman's "naked eye," visual signs of good health and ill health are usually easy to spot. For years evolutionary psychologists have told us that we're wired to be pretty adept at assessing the health and, by extension, fertility of the opposite sex. This, of course, is important for purposes of procreation and survival. What we see as health in another person is interpreted as attractiveness. These same psychologists have also shown through studies that we judge people's attractiveness based more on proportions than on weight—as long as they aren't emaciated or severely obese. For women the ratio of the waist to the hips is important, and for men the shoulders figure into it. Different cultures may deem certain body types more attractive than other types, but patterns even across disparate cultures do exist that play into universal images of health. And while certain proportions may signal something about overall health, that health cannot be measured with the current body mass index.

A Convoluted Debate with a Simple Prescription: The 10 Key Principles

Most commonly accepted rules for good health (e.g., don't smoke; drink alcohol in moderation; and break a sweat once in a while) can be quite exacting as to how to follow or implement in one's life. As you can see, the obesity paradox's lessons and applications are much less exacting and one-size-fits-all—which is why I cannot tell you precisely what you should be eating for dinner, how many pounds is too many, or when to lose an inch off your waist. I'm also not here to diagnose you. Instead, I'm here to empower you to take control of your body and the future of your health. The suggestions I offer in the following pages are more like lifestyle algorithms—guidelines for thinking through your myriad lifestyle choices and personal circumstances in light of all this information. And of course, the choices you ultimately make must be tempered by your values and individual codes of ethics and behavior.

Principle #1: Put Obesity into Personal Perspective

First, let me remind you: The facts of the obesity paradox do not tacitly endorse obesity to any degree. From a public health perspective, preventing weight gain in the first place should be the primary goal. As many can attest, sustained weight loss is challenging; and the more you gain beyond an ideal weight, the harder permanent weight loss becomes and the more you place yourself at risk for numerous chronic conditions that will be further aggravated by the extra weight and can impact your quality of life. I do not intend all my musings on the potential benefits of fat and misunderstandings about what it means to be obese to give people permission to pay little attention to their weight.

More than anything, my hope is that my message empowers you to see your personal issues with weight under a new light that can inform better decisions about maintaining the weight that's ideal for you and your risk factors. If you have a history of diabetes or heart disease in your family and already show signs of metabolic dysfunction, then you know that you would do well to keep a closer eye on your weight and to stay as fit as possible. But even then, you probably don't need to "lose those last ten pounds" or set your sights on looking like a model; our culture has a way of idolizing low-fat body types that are on the extreme side. It also has an insidious way of making you think that diet x, y, or z will save your health and that you'd be better off if you were several pounds lighter. But that may not be the case. Remember what I pointed out earlier: High-quality studies show that obese individuals who are metabolically normal don't improve their health with weight loss. Some of us may be genetically prone to metabolic conditions regardless of weight, just as we are each probably genetically encoded to be at a certain weight.

So the old logic rings true: Preventing weight gain and obesity will go a long way toward protecting your health and reducing your chance of getting certain diseases, much less dying from one of them. But as I've also addressed throughout this book, the term *obesity* is deserving of a new definition. The BMI may say that you're overweight or obese, but if you're fit at the same time, then there's much less to worry about (assuming you steer clear of morbid obesity). And as I've also proven (hopefully) by now, controlling your weight through diet alone can have some unintended, horrible consequences from both a metabolic and a cardiovascular standpoint. Mind that U-curve and find your BMI sweet spot, which can be anywhere between 18.5 and 35—so long as you maintain fitness. People with naturally larger body frames will have higher BMIs, and that's okay if your metabolic health remains good and any chronic condition is carefully managed.

Based on all the current science to date, I'm comfortable saying that 27 is the new 23. And what I mean by that is if your BMI says you're overweight or even mildly obese, you don't need to focus on losing weight if you're fit (and if you don't know if you're fit, see Principle #9). If your body weight is considered normal and you're fit, stay that way. Don't try to gain weight, but be sure not to lose your fitness. If the choice is between sacrificing slenderness or fitness, it's best to remain fit regardless of weight gain. It begs to be reiterated: Being thin and out of shape is worse than being overweight/obese and fit.

Principle #2: Change the Conversation

Although we like to think that we each have a pretty good idea whether or not we should lose a few pounds in the name of health (not vanity), this is virtually impossible to know due to the prevailing discourse about fatness and society's love of thinness. How much have your thoughts about your weight affected your quality of life and mental health? How many times have you tried a diet, only to lose weight and then gain it all back, plus more (and then felt terrible about yourself)? How often do you wish you weighed less? If you had to give a number between 1 and 10 as to how much your weight factored into your happiness (1 being not at all and 10 being a lot), what would you say? My guess is many of you would say a number above 5. Unfortunately, I don't expect our obesity-loathing environment to change anytime soon, which means we must each first alter our internal conversation to arrive at our own definition of health.

The conversation needs to shift to one that centers on metabolic fitness, which includes the cardiorespiratory component. I'd rather tell an overweight or obese patient to work on improving his or her fitness level than to focus on severely cutting back calories and dramatically changing eating habits overnight. Not only does this approach have a much more positive spin to it, but it tends to have the effect of promoting weight loss without all the stressful, abstemious aspects of the thought of "going on a diet" for the sole purpose of losing weight. This isn't to say that nutrition isn't as important as exercise, however. I want my patients to achieve a state of health that's measured by how well their metabolism functions and how strong their cardiorespiratory system is—not by how much they weigh and what size dress or pants they wear.

Doctors who push weight loss have good intentions, but I worry that the message people hear is that they should lose weight at whatever cost—trying extreme diets and controlling their weight through eating habits alone. That can be hard on a body, as you know by now, unless the fitness component is addressed. People who deal with chronic conditions are particularly vulnerable, and would benefit more from attention to physical fitness than solely dietary choices. So indeed, doctors should be advising that their obese and overweight patients aim to be as fit as possible despite their weight, but the conversation cannot end with just a directive to "lose weight." If weight loss is prescribed, then it should be encouraged via better fitness. All roads lead to ideal weight once fitness takes center stage.

The idea that we need to alter the ongoing conversation about fat from one about weight to one about health was recently made all the more apparent to me upon reading about the so-called "200-pound anorexics"—obese teens who will one day struggle with anorexia nervosa because they keep hearing the message that fat is bad, and as such, they end up going to extremes to lose it. As a society, promoting weight loss to combat obesity is well meant, but the antiobesity movement can have unintended devastating consequences. Taken too far, it can result in focusing on thin versus fat, instead of healthy versus unhealthy. And indeed, teaching habits such as counting calories or avoiding carbs or calling this food "good" and that food "bad" can all too easily slip into the obsessive patterns associated with eating disorders. And such disorders don't just affect teens. They can prey upon adults, too. An estimated thirty million Americans will have an eating disorder sometime in their life.

Sadly, society perpetuates the idea that any weight loss is good for an overweight or obese person, no matter what. But we fail to ask ourselves: How is that person losing weight? Is he going to extremes, such as by going on a crazy, unsustainable diet? Is he doing what I've already said is perhaps the worst: controlling his weight (and weight loss) through dietary restrictions at the expense of fitness?

Let's change the conversation. As we try to partly turn a blind eye to how BMI defines us, let's also evict the word *fat* from our vocabulary and replace it with *health*. As I've already proven, you can be remarkably healthy at many different BMI values.

Principle #3: Know Your Numbers

In addition to having a vague idea of where you are in the wide range of health possibilities, one of the best ways to measure your health more precisely is to have your doctor check your vital signs and order a complete metabolic and lipid panel from a simple blood draw after you have fasted for twelve hours. It will include blood counts, values for certain proteins and electrolytes, markers of inflammation, some hormonal values, vitamin D levels, and cholesterol and triglyceride levels. You'll want to do this once a year at your annual checkup. The results to pay attention to include the following:

- **Hemoglobin A1c:** Hemoglobin A1c (also called *glycosylated hemoglobin*) is the protein found in red blood cells that carries oxygen and binds to blood sugar, and this binding is increased when blood sugar is elevated. Hence, the test determines your blood sugar levels, and whether or not you're at risk for diabetes or already diabetic. While hemoglobin A1c doesn't give a moment-to-moment indication of what your blood sugar is, it is extremely useful in that it shows what your "average" blood sugar has been over the previous ninety days. This is why hemoglobin A1c is frequently used in studies that try to correlate blood sugar control to various disease processes ranging from diabetes to coronary artery disease and dementia. An ideal hemoglobin A1c would be in the 4.2 to 5.6 percent range. Values of 5.7 to 6.4 percent indicate an increased risk for type 2 diabetes; values greater than or equal to 6.5 percent mean you have type 2 diabetes. You can lower your hemoglobin A1c number and improve insulin sensitivity through dietary shifts (e.g., reducing your carbohydrate consumption) and physical exercise. In one notable study published in 2010, researchers told thirty participants to make no lifestyle changes, while putting thirty-five others on an exercise program three days a week. The control group did not participate in any form of exercise. After the sixteenth week, hemoglobin A1c decreased by 0.73 in the exercise group but increased by 0.28 in the non-exercise group. To put these numbers in context, if your hemoglobin A1c was 6.0, a reduction of 0.73 represents a 12 percent decrease, which rivals diabetes medications.

• **Fasting blood glucose:** Your in-the-moment blood sugar level tells you how much glucose is currently circulating in your blood, which can be an indicator of your risk for diabetes or that you're currently diabetic. You want this number to be ideally less than 100 milligrams per deciliter (mg/dL), or at least less than 110 as I discussed previously. (Values of 126 and higher indicate diabetes.)

• **C-reactive protein (CRP):** This is a marker of inflammation. Higher levels are correlated with higher risk for a variety of illnesses and metabolic conditions, including diabetes, heart disease, and obesity. This number should be between 0.00 and 2.0 mg/L (ideally, less than 1.0).

• **Lipids:** These include your fasting cholesterol numbers (total cholesterol, HDL, and LDL values) and your triglyceride values. A desirable total cholesterol is between 120 and 200 mg/dL, but this can be offset by a healthy ratio of good to bad cholesterol (more good, HDL cholesterol and less bad, LDL cholesterol). In other words, if your total cholesterol is high but you've got high good cholesterol and low bad, you're probably okay. An ideal HDL level is greater than 60 mg/dL, but levels above 40 for men and above 50 for women are decent. An optimal LDL level is less than 100 mg/dL. Anything between 100 and 129 mg/dL is borderline, 130 and 159 mg/dL is considered mildly high, 160 to 189 is deemed moderately high, and an LDL level of 190 and above is severely high. A target triglyceride level is less than 150 mg/dL, with values between 150 and 199 being borderline, 200 to 499 being elevated, and values 500 and higher being severely elevated, placing one at high risk of the potentially deadly disease, pancreatitis, as well as heart disease.

I recommend that you keep track of your blood pressure periodically throughout the year to note changes. You can buy a blood pressure monitor to keep at home or find a local pharmacy or drugstore that offers free tests. Although I don't want anyone to become addicted to checking his or her weight on the scale, we do know from numerous studies that tracking our weight on a regular basis can help us to maintain it. What do I mean by "regular basis"? That's up to you, but in the least aim to check your weight once or twice a month (not daily). You should weigh yourself at the same time of day, ideally in the morning before you've eaten breakfast.

Principle #4: Improve Your Metabolic Health

Knowing your numbers will help you focus on other areas where there's room for improvement. When patients ask me how they can enhance their health without going on a traditional diet or even thinking about weight loss, I share the following five tips:

- *Reduce sugar consumption.* This will automatically help you to gain better control over your blood sugar balance. Most everyone—regardless of blood sugar levels—would do well to heed this advice, since imbalances can start far under our conscious radar, secretly pushing us toward metabolic mayhem. And of all the ways in which we can limit the amount of sugar we consume, the most effective strategy to implement right away is to limit sugary beverages (e.g., soda, sports or energy drinks, fruit juices) and refined carbohydrates, sugars, and grains found in numerous processed food products. In the beverage department, choose low sugar alternatives or stick with just water. As you move away from sugar-laden foods, you'll gravitate toward whole grains, vegetables, and lean proteins—all of which contain more nutrients and compounds, like fiber, that support healthy blood sugar. You don't have to nix all sugar, however, especially if you're highly active. Small amounts of sugar or an occasional sugary beverage won't kill you or significantly increase your risk of anything so long as you engage in regular physical activity and exercise.

• *Maintain regular sleeping patterns.* This is not a trivial point (particularly if you're sleep deprived, as many of us are today). Both laboratory and clinical studies have shown that virtually every system in the body is affected by the quality and amount of sleep we get at night. In addition to sleep's well-known benefits, such as helping us to feel refreshed, creative, happy, and mentally sharp, it can influence our appetite, how fat we get, whether we can fight off infections, and how well we can cope with stress. It does this by playing a commanding role in orchestrating our hormonal balance. As I've already described, our hormonal balance affects a wide variety of processes in the body that are part of our overall health, especially when it comes to our metabolism. People who keep a normal sleep schedule daily (to the best of their ability) are typically healthier by every measure and have an easier time maintaining a healthy weight and fitness. So if you can't say that you get a good night's sleep routinely, then make this an area to address in your life (visit www.DrChipLavie.com for resources that can help you achieve this). Aim to go to bed and wake up at the same time daily, weekends and holidays included, and make sure you're getting enough sleep; most of us need seven to nine hours a night. As I covered in chapter 7, losing sleep will throw your appetite hormones out of whack: Several studies over the past decade have shown that sleep-deprived individuals suffer a drop in leptin, which causes them to have a whopping 24 percent increase in hunger and appetite. Moreover, they crave foods that send their blood sugar soaring.

- *Reduce exposure to obesogens.* Choosing foods that are less processed and closer to nature (i.e., they don't come with a Nutrition Facts label) will help you reduce your exposure to environmental toxins that could be having a negative impact on your body's physiology, especially your metabolic hormones. Other ways to limit contact with these chemicals and compounds include buying BPA- and phthalate-free plastics, canned goods, and products (you can search the Environmental Working Group's database at www.ewg.org to find out where these substances lurk), and buying organic foods and products whenever possible. I don't expect you to suddenly triple your grocery bill, but be more conscientious about what you spend money on. You can often find healthy trades: Skip the boxes of processed snack food (e.g., chips, crackers, and cookies) and find healthier alternatives (e.g., whole fruits, nuts, and nut butters); swap your grain-fed chicken and processed meats for wild fish and grass-fed beef. Opt for antibiotic-free poultry. Cook with healthy fats like extra virgin olive oil rather than butter, margarine, and hydrogenated oils. These may seem like expensive alternatives, but when you add up how much you likely spend on processed, packaged foods as well as fast food, you might find that you will break even.

- *Move more throughout the day.* High fitness, one of the core ingredients to a healthy metabolism regardless of weight, needn't be acquired through formal workout sessions lasting an extended period of time. Nor does it mean engaging in a rigorous routine on a par with training for an athletic competition. I'll outline my recommendations for achieving "high fitness" in Principle #9; for now, however, just be mindful that you can increase your fitness all day long just by moving as much as possible. If you have a desk job, make an effort to get up and move for a few minutes at least once an hour. Take your calls on a wireless device so you can walk around your office. Use twenty minutes of your lunch break to go for a brisk walk. Avoid elevators when you can and choose the stairs. Park at the far end of lots. Be creative in seeking ways to move your body despite typical obligations that would have you remain sedentary. You can be active today without having to join a gym or even put on traditional workout clothes. Get innovative. And get off your butt.

Principle #5: Maintain Dietary Habits That Work for You

The best dietary protocol is the one that works for you, period. It's beyond the scope of this book to lay out a single regimen with meal plans and menu ideas. Note that a "diet" doesn't have to entail a regimen geared toward weight loss—and for some, this way of thinking is not the ideal approach (refer back to Principle #4). In 2013, researchers from Duke University found that by focusing on not gaining weight, rather than losing ("maintain, don't gain"), people are more likely to stay committed to a long-term plan and avoid packing on those one or two pounds a year that add up over time. Such "weight creep" is part of what pushes someone beyond the overweight boundary and into the obesity realm, where more serious health challenges and risks can lie in wait.

Whether you choose to go gluten-free, low-carb, vegan, or raw or join a program like Weight Watchers is totally up to you. It doesn't really matter as long as you reach your health goals and, above all, enjoy what you're eating and your body seems to agree. Even many of today's popular diets that are geared toward weight loss can be used simply to teach you healthier eating habits. They don't have to be about weight loss, but they are often chock-full of wise ideas about controlling metabolic health through a tasty array of food choices while giving you fresh recipes. Just be sure that you don't force yourself to adhere to an impossibly strict protocol or the latest fad that will eventually backfire, leading you back to your old unhealthy ways. Remember, just as there are many religions in the world, there are many healthy eating traditions, all of which follow similar patterns, like encouraging moderate portions, favoring foods that support metabolic health, cooking, letting hunger build in between meals, and regarding treats as what they are—treats. Any traditional diet is better than our processed food culture, and traditional eating habits have worked for centuries among different peoples with vastly different diets around the world.

Individual risks and needs must be taken into account. For those with diabetes and/or metabolic syndrome or insulin resistance (characterized by abdominal obesity, high triglycerides, low HDL, and elevated glucose), a diet lower in sugar and carbohydrates works well. But if you're engaging in high amounts of physical activity, you'll need more carbs and even simple sugars. Of course, it always helps to choose those we'd all rank higher on our heath meters—carbs from natural sources such as fruits, vegetables, and whole grains, rather than processed varieties.

Principle #6: When in Doubt, Go Mediterranean

In recent times, we've heard a lot about the benefits of the Mediterranean diet, which is famous for being rich in olive oil, nuts, beans, fish, whole grains, fruits and vegetables, and even wine with meals. It endorses a variety of plant foods while reining in red and processed meats and high-fat dairy. It also encourages the use of herbs and spices instead of salt and unhealthy fats to make foods tasty. In March 2013, *The New England Journal of Medicine* published a landmark study showing that people age fifty-five to eighty who ate a Mediterranean diet were at lower risk of heart disease and stroke—by as much as 30 percent—than those on a typical low-fat diet. They also had a reduced risk of death from cancer, as well as a reduced incidence of Parkinson's and Alzheimer's diseases.

The Mediterranean diet isn't necessarily a weight loss diet. It's simply a way of eating that can indeed lead to weight loss, but overall it reflects a healthy food lifestyle with documented health benefits. In fact the Dietary Guidelines for Americans recommends the Mediterranean diet as an *eating plan*—not a diet per se—that can help promote health and prevent disease. So if you're struggling with knowing exactly what to eat to support a healthy weight, you may want to check this one out. The beauty about the Mediterranean diet is that it's one your entire family can enjoy and there's plenty of wiggle room for occasional splurges. And unlike so many other "diets" on the market, it also recognizes the importance of being physically active. (For more on this, go to the Mayo Clinic's website and search for "Mediterranean diet" or check out www.oldwayspt.org, a nonprofit food and nutrition organization that endorses this diet.)

Principle #7: Don't Buy into Myths if You Choose to Go on a Diet to Lose Weight

Myths about weight loss and even obesity continue to circulate like urban legends. They are what urged my esteemed colleague Dr. David B. Allison, who directs the Nutrition Obesity Research Center, at the University of Alabama at Birmingham, to take a closer look and see what the data actually says. His report, published in early 2013 in *The New England Journal of Medicine*, aims to eliminate unproven assumptions that are repeated so often and with such conviction that they delude even doctors and scientists.

I realize that some of you will be steadfast about dropping a few pounds no matter what. If so, first be sure that you're doing it for the right reasons—for instance, to feel better about yourself, to engage in activities that your weight currently prevents you from doing, to reduce your inherited risk of heart disease. Then make sure that you're mindful of the most prominent myths and truths that Allison and his colleagues revealed. These will help you to reframe the word *diet* in your head and hopefully approach your efforts with patience and realistic conviction rather than frustration and anxiety.

Most of us have heard that most diets fail. We've also heard things like "walking a mile a day will melt heaps of fat off," "formula-fed babies are doomed to be fat," "crash diets are unhealthy," and "vending machines and lack of phys-ed classes in schools make kids obese" (remember the correlation vs. causation lesson?). We fall for these ideas, many of which are unproven assumptions, not out of stupidity but probably because we hear them repeated so often. And many of them seem fairly reasonable at face value. But no sooner did Allison delve into the obesity and weight loss literature than he discovered the power of "reasonable bias"—our tendency to accept what sounds so reasonable that it must be true.

Take, for instance, the idea that you're most likely to succeed on a diet if you have a realistic weight loss goal in mind. Sounds reasonable, right? But in reality, based on reams of data, that's not always the case. Allison found no consistent association between people's intentions or the ambitiousness of their goals and how much weight they lost and how long it stayed off. But such a myth illustrates the challenge for weight loss programs: How can they sell a promise when the science can't prove—or guarantee—results? On average, people don't actually lose all that much, maybe a maximum of 10 percent of their total weight. As it turns out, setting 10 percent as your target weight loss goal, however, might not be the ideal strategy. And as I just mentioned in Principle #5, setting any weight loss goal might not be best. Just focusing on not gaining could be the way to go. Let's take a look at some of the other myths Allison uncovered, for their truths may help you to find a healthier way to diet.

> *Myth:* Small activities make a big difference. For example, walking a mile a day can lead to a loss of more than fifty pounds in five years.
>
> *Truth:* Although one can argue that every minute engaged in exercise counts (and counts toward abiding by the half-century-old "3,500-calorie rule," which states you only need to create a deficit of thirty-five hundred calories to lose one pound), recent studies have shown that there's enormous variability from person to person, especially once the body undergoes changes through diet and exercise. While it's nice to think that, according to this rule, you can increase your daily energy expenditure by a hundred calories by walking one mile per day and lose more than fifty pounds over a period of five years, your true weight loss will prob-

ably be only a fraction of that—about ten pounds (assuming you maintain the same caloric daily intake, too!). Why? Because changes in body mass also change your body's energy requirements. It often takes bigger, more ambitious activities to stimulate dramatic weight loss, especially weight loss that's sustainable as the body changes. (A landmark study is currently under way by Drs. Gregory Hand and Steven Blair at the University of South Carolina that shows changes in metabolic rate depending on whether one is on the positive or negative side of the "energy balance" curve.)

Myth: Setting a realistic goal to lose a modest amount works. People who are too ambitious will give up quickly out of frustration.

Truth: You can set an unrealistic goal (say, losing forty pounds in six months) and still experience great results, as several studies have shown that more ambitious goals are sometimes associated with better weight loss outcomes

Myth: You have to be mentally ready to diet or you will never succeed.

Truth: Being "ready" does not predict how much weight you'll lose or how well you can stick with a program you're trying to follow. The explanation, according to the researchers, may be simple: When we voluntarily choose to enter a weight loss program, we are already at least minimally ready to engage in the behaviors required to lose weight.

Myth: Slow and steady weight loss works best. If you lose weight too fast you will lose less in the long run.

Truth: A large analysis of randomized controlled trials showed that this simply isn't the case. Whether you lose weight slowly on a semi-ambitious diet or go on a more restrictive diet that has you losing weight quickly, both types of diets can work in the long run. You just need to do what works for you.

Among some of the other myths debunked by Allison and his team are that physical education classes play an important role in reducing or preventing childhood obesity; breast-feeding protects kids against obesity; and sex is a calorie incinerator.

Indeed, as counterintuitive as it sounds, physical education has not been shown to reduce or prevent obesity. Three different studies indicated collectively that even when schools increased the number of days children attended physical-education classes, it didn't have much of an effect on their BMIs. And two large-scale analyses further showed that even specialized school-based programs that promoted physical activity didn't put a dent in reducing BMI or obesity. No doubt there has to be a level of physical activity (which the researchers define as a specific combination of frequency, intensity, and duration) that's effective in combating obesity, but the question remains: Is that level achievable in conventional school settings? Nevertheless, whether obtained before, during, or after the school day, clearly increasing physical activity is needed for children and adults alike. Long-term increases in physical activity and exercise may be even more important to maintaining fitness and fighting off many diseases rather than preventing or treating obesity per se.

For years now we've been told that in addition to breast milk being best for babies for a variety of health reasons, breast-fed babies are less likely to suffer the ills of obesity later in life. But surprise—the idea that breast milk is protective against obesity may not be true. Several studies involving thousands of children have provided no compelling evidence of an effect of breast-feeding on obesity. And it's been disclosed that even the World Health Organization's report stating that people who were breast-fed as infants are less likely to be obese later in life was seriously flawed. The report reeked of unscientific biases and failed to rule out confounding factors that could have influenced the association made between breast-feeding and risk of obesity. One of the reasons this myth persists is because breast-feeding does have important potential benefits for both mother and child that have nothing to do with obesity risks, so health professionals don't want to discourage breast-feeding to any degree. This is the same mentality we've seen in the obesity arena: Despite the data, many health professionals are loath to admit to patients that some forms of obesity are okay.

And I hate to break it to you, but a bout of sexual activity isn't equivalent to running a few miles. I love how one of the authors of Allison's study phrased it: "Given that the average bout of sexual activity lasts about 6 minutes, a man in his early-to-mid-30s might expend approximately 21 [calories] during sexual intercourse. Of course, he would have spent roughly one third that amount of energy just watching television, so the incremental benefit of one bout of sexual activity with respect to energy expended is plausibly on the order of 14 [calories]."

UNPROVEN IDEAS

- Diet and exercise habits in childhood set the stage for the rest of life. Although your BMI tends to follow a certain pattern as you age, and remain steady in a similar percentile range, longitudinal genetic studies suggest that such patterns may be due to genes rather than early life experience. Surprisingly, we don't have any sound science from clinical trials to provide evidence to the contrary.

- Eating breakfast is key to weight loss and protective against obesity (I think every diet book I've read preaches, "Never skip breakfast"). But two respectable studies that looked at the difference between eating breakfast and skipping it showed that it didn't matter when it came to weight.

- Adding lots of fruits and vegetables to your diet will help you lose weight or not gain as much. Yes, eating fruits and vegetables has health benefits. But when you don't change other dietary habits that could be sabotaging your good intentions, more fruits and vegetables will not necessarily prevent weight gain.

- Yo-yo dieting is harmful and can lead to increased death rates. Although we have some evidence to show that fluctuations in weight are associated with increased mortality, such findings are probably due to other health-related conditions and not the yo-yoing itself.

- Snacking is bad for the waistline; people who snack gain weight and get fat. Maybe not. Numerous clinical trials do not support this presumption, and there's no scientific evidence to show a consistent link between snacking and obesity or increased BMI. Eating more nuts may be helpful for general health and for prevention of weight gain, probably by reducing hunger and cravings for higher caloric food choices.

- If communities focused on access to physical activity and infrastructure, such as by adding bike paths, jogging trails, sidewalks, and parks, people would not be as fat. Unfortunately, we just don't know if such conclusions can be drawn. While it's common sense to think that our physical environment can impact our weight, we don't have scientific proof of such a seemingly obvious association. But we do know that physical activity burns calories and increases metabolic rate, which should produce benefits. So to a large extent we have to create the best environment for ourselves that we can to motivate us to move.

Principle #8: Know the Facts About Weight Loss

Given the laundry list of diet myths, you'd think that the list of truths backed by scientific facts would be equally long, if not longer. But the "truth" is much simpler, and by now you probably already know it somewhere deep in your subconscious:

- Heredity is important but it's not destiny. So even though you may come from a family with a history of diabetes and heart disease, for instance, these conditions do not have to be in your fate or severely infringe on your quality of life. You can do a lot through lifestyle choices to change not only your risk factors but also how much of an impact certain conditions have on you if you do develop them.

- Exercise helps with weight maintenance. I will add that exercise does a lot more than that. It can help you to control blood sugar, stave off diabetes, and inhibit the negative effects that carrying extra fat can have on you. Most important, physical activity is the best way to maintain and improve your overall fitness, perhaps the most potent of all the risk factors for heart disease and overall health (see Principle #9).

- Weight loss is more successful with programs that provide meals. If you do intend to lose weight for whatever reason, then be sure to choose a program that includes plenty of meal ideas so you're not left wondering what to eat. Weight loss is hard enough; you don't need to add to that difficult task the chore of figuring out exactly what to cook or eat.

- Some prescription drugs help with weight loss and maintenance. I will temper this statement by adding that no weight loss drugs replace the need to make healthy changes in eating habits and activity level. And they are often reserved for those with a BMI greater than 30 and for people who have serious, uncontrolled medical problems related to obesity.

- Weight loss surgery in some patients can lead to long-term weight loss, less diabetes, and a lower death rate. As with prescription weight loss drugs, surgical options are often suggested only for severely and morbidly obese people who have failed to lose weight through diet and exercise.

Although some people might wish some of the myths about weight loss were indeed true, here's the (truthful) silver lining: If we look at all the science about the overwhelming task of losing weight permanently, it turns out that the journey to an ideal weight might not be that challenging after all. This is a conclusion shared by Allison and his colleagues based on his study. Some of the most dreadful myths that people cling to are the most destructive to any attempt to lose weight, so once you rid yourself of these false thoughts, effortless and healthy weight loss will suddenly become more attainable.

Principle #9: Prioritize and Maintain Fitness Regardless of Weight

Not everyone can maintain their high school waistline as they age, and as I've already proven, that's okay so long as you stay fit. Fitness is much more important and will help you to live a long, robust life despite the dreaded weight creep through the years. It's so important that in 2013 I coauthored an article on behalf of the American Heart Association (published in the journal *Circulation*) that stressed the importance of cardiorespiratory fitness. Acting as a policy statement led by my close friends Lenny Kaminsky and Ross Arena, the piece called for a national registry in the US to serve as a database for tracking health behaviors and risk factors about this critical element of health. As we stated in the article, my colleagues and I feel that the importance of cardiorespiratory fitness has often been neglected in health circles, especially in our area regarding risk for heart disease. And unfortunately, fitness levels remain the only major factor that is not routinely and regularly assessed in clinical settings.

I'm pretty certain that if all the studies on the obesity paradox alone included the fitness component, we'd see that it's longevity's pièce de résistance despite weight. In my studies that do consider the fitness component, I've seen over and over again how critical it is to avoid low levels of fitness for your age and sex, and the biggest bang for your buck occurs when moving from the bottom level of fitness to the next rung up the ladder. In other words, you gain a lot from a health standpoint when you move from the bottom 20 percent to the 20-to-40 percent range (though you're still below average at this level). Your risk is further lowered by moving to the 40-to-60 percent range (average), and then again to the 60-to-80 percent range (above average), and finally to the most fit category of people in the top 20 percent. The closer you get to the fittest group of people, however, the smaller the leaps made in terms of health benefits. You stand to gain more health benefits by moving from below average to just average than from average to above average. What this really means is you don't have to achieve top levels of fitness to reap tremendous health rewards.

How fit are you now? And how can you know what your fitness level should be, given your age and gender, so that you're not in the bottom 20 percent? In general, if you can climb several flights of stairs without difficulty and can walk a mile or two at a decent pace (e.g., a mile within fifteen minutes), then you are in good shape at any age, whether you're a man or woman. There's usually room for improvement, however. (Studies that compare different fitness levels between men and women as well as among people of varying ages note two facts: [1] women in general have lower fitness levels as compared to men due to their physiology, and [2] as you age, your cardiorespiratory fitness level will naturally decline as heart rate lowers during the aging process.)

In terms of muscle mass, you can probably just take a good look at yourself in the mirror to see if you've got better-than-average muscle tone. And you've probably got some decent muscle strength if you can complete your normal daily activities, including climbing stairs and lifting heavy objects like children or grocery bags, without much strain. (In studies that investigate the effects of muscular strength on various health parameters, researchers typically use things like handgrip strength, a standing long jump, push-ups, and sit-ups to figure out a muscular fitness score.)

You don't need to sign up for a 10K or take up Spinning to attain "high fitness" (and if you are an exercise fiend worried about overdoing it, see chapter 9). A well-rounded exercise program that builds and maintains high fitness includes cardio work, strength training, and stretching. Each of these activities confers unique benefits that our body needs for peak performance and to positively affect our genes and metabolism. Cardio, which gets the heart rate up for an extended period, will burn calories, decrease body fat, and strengthen both the heart and lungs; strength or resistance training will keep your bones strong and prevent loss of lean muscle mass; and stretching will keep you flexible and less susceptible to chronic inflammation. All of this exercise will support metabolic health too, and help you increase your body's sensitivity to insulin (which, you'll recall, is a good thing to help avoid diabetes).

I recommend that you aim for thirty to forty minutes of aerobic exercise on most days of the week (at least four to five times weekly is usually sufficient). That means engaging in an activity that gets your heart rate up at least 50 percent from its resting rate. This can be done through brisk walking, jogging, biking, stair climbing, swimming, using an elliptical machine, or taking an aerobics class. In addition, try to work your big muscle groups twice weekly with resistance training by using light weights or gym equipment, or via classes geared toward this goal (such as Pilates and certain forms of yoga). If you're using classic barbells, which you can buy cheap and keep at home, here's a simple rule to follow: Lift enough weight to complete twelve to fifteen repetitions without considerably straining yourself. If you can do endless repetitions, you need to increase the weight to gain the health benefits.

Keep in mind that the benefits of exercise are, for the most part, cumulative. You can engage in short bursts of exercise throughout the day (which can actually help minimize your time spent sitting), or commit to a routine that blocks out an hour or so for your workouts. Just be sure that if you do dedicate a single period of time to your exercise regimen, you don't allow yourself to be sedentary the rest of the day. Even just breaking up your sitting time by walking around with a pair of free weights and doing a few biceps curls can lower your risk of disease and premature death.

Now that you have the proper knowledge, you can determine your own exercise training routine, and how it applies to your goals. As always, I recommend you seek medical help to make sure you are healthy enough for physical activity before beginning any new exercise routine—especially if you're being sedentary and have cardiovascular risk factors. And there are plenty of nifty devices and apps (some of which you can download for free on your smartphone) to help you track your movement and exercise habits. For a list of ideas, go to www.DrChipLavie.com.

Principle #10: Consider Drugs and Supplements as Needed

It seems like we are bombarded by the media daily regarding the use of drugs and supplements, with messages about what we should be taking to "boost metabolism," "spur weight loss," and "enhance health." One day it's reported that certain medications or vitamins are good for us and will extend our life; the next, we read about how some can do more harm than good, increasing our risk for certain diseases, including cancer. While it's true that drugs and supplements should never be used as an insurance policy against lapses in our diet and lifestyle, there are specific products proven to work in supporting our health. Let me give you my thoughts on the most commonly used ones.

Daily multivitamin: If you eat a balanced diet, you shouldn't need to worry about nutrient deficiencies. It's up to you (and your doctor) whether or not you choose to take a daily multivitamin. Although many people swear by their multivitamins and won't want to give them up, some recent studies are actually showing that they may not be as beneficial to health—potentially increasing your risk for certain cancers. The lesson here is that you can't expect to take a pill to satisfy nutritional needs that real food provides.

Statins: Depending on your cholesterol levels, age, and risk factors, your doctor may encourage you to take a statin, a compound that inhibits a liver enzyme that plays a central role in the production of cholesterol. This is a decision to ultimately make with your doctor's guidance.

For a long time we thought the purpose of statins was solely to lower heart disease risk, by virtue of their impact on cholesterol production. But the story doesn't end there. It turns out that they have a profound effect on the entire body, because they help control systemic inflammation—the biological process that can go into overdrive and trigger all kinds of dysfunction and illnesses when left unchecked. You'll recall that when a body has high levels of inflammation markers, it means that it's encountering harmful stimuli, from germs to damaged cells to irritants. Inflammation is a survival tool, but in excess—just like exercise and fat—it can be harmful, hence the connections made between certain kinds of inflammation and our most pernicious and chronic diseases, from obesity to cancer, as well as an accelerated aging process in general. Studies in the past decade show that statins can reduce the risk of first-time heart attacks, strokes, and even death from heart disease by about 25 to 35 percent (and in many of these people, high cholesterol is not a problem). Studies have also shown that statins can reduce the chances of recurrent strokes or heart attacks by about 40 percent. In 2012, *The New England Journal of Medicine* published a study involving three hundred thousand people that indicated a dramatically lowered risk of death from cancer among those who took statins.

Not everyone can tolerate statins, which have side effects, most notably muscle aches and weakness. For patients who would benefit from a statin but cannot tolerate the muscular side effects, I recommend 200 milligrams per day of coenzyme Q10 (CoQ10), an over-the-counter supplement that may help and has been suggested by some to help heart failure. Your body naturally makes CoQ10, a substance similar to a vitamin that cells use to produce energy. It also helps enzymes perform certain bodily processes, thereby protecting the heart and skeletal muscles. (In 2012, I published a paper with my colleague Dr. Richard Deichmann, in which we showed mild improvements in aerobic and muscle performance among athletes over age fifty who were on statins when 200 mg of CoQ10 daily was added to their regimen.) Additionally, low levels of vitamin D can increase muscular symptoms, with or without statins, which can be helped or eliminated with vitamin D replacement.

Omega-3: For people with high cardiovascular risk and/or known heart disease, I recommend taking an omega-3 supplement (EPA/DHA combination) at a dose greater than 600 mg per day. In general, people who don't typically consume a lot of fatty, cold-water fish, such as wild king salmon, arctic char, and black cod, would do well to consider an omega-3 supplement. Research shows strong evidence that the omega-3s EPA and DHA can help lower triglycerides and improve vascular function; recent evidence suggests that people not on statins may gain even greater value with omega-3 supplements compared to patients who are already well protected by statin therapy. Again, the pros and cons of any supplement should be determined with one's physician.

Q. Fish oil supplements have been both praised and persecuted by health experts lately. What is the verdict on whether or not these are helpful to take for heart health, among other benefits?

A. As soon as the *Journal of the National Cancer Institute* published a study in 2013 suggesting a higher risk of prostate cancer among men who had higher blood levels of omega-3, my colleagues and I sprang into action. We noted several issues in the study. For one, the results showed an association and not causation. And second, the "high" blood levels were much lower than levels obtained by eating large quantities of fatty fish or taking omega-3 supplements. Therefore, this study should not have concluded in a blanket statement to the public that fish oil supplements cause prostate cancer. In a paper that we quickly published in another journal (and followed by others), we argued that omega-3 fats (specifically, DHA and eicosapentaenoic acid, or EPA) are not associated with prostate cancer and may actually be associated with a lower risk of the disease (research has long shown that these omega-3s have an anti-inflammatory effect, which is protective against cardiovascular problems and reduces the risk of many types of cancer). People with prostate cancer in particular seem to do better with more omega-3s in their diet. The problem with recent research is that we have fewer people to include in studies showing the protective effects of omega-3s—so many individuals are benefiting from taking things like baby aspirin and statins that there aren't as many people having cardiovascular events in general. I suspect that this study's findings will be confirmed by others or that fish intake or omega-3 supplement usage will be linked to prostate cancer. And, as the investigators themselves say, the potential cardiovascular benefits outweigh the unlikely prostate risks. (For more particulars about this debate, see www.DrChipLavie.com.)

That said, taking omega-3 supplements is no substitute for keeping your blood pressure and blood fats within the healthy zone and engaging in the kinds of activities good for the heart and body. Maintaining a healthy diet and exercise will do more for your preventive program than any supplement you take.

Vitamin D: If your lab results indicated that your vitamin D levels are low (probably anything below 30 nanograms per milliliter and certainly below 20), you may want to consider taking a daily vitamin D_3 supplement (1,000–2,000 IU daily; folks with extremely low levels may need a much higher dose that requires a prescription), which I've found to be very effective. As I detailed earlier, vitamin D is not really a vitamin; it acts more like a steroid or hormone in the body. Vitamin D deficiency is not just about an increased risk for weak, soft bones and, at the extreme end, rickets; it's associated with many conditions, including diabetes, obesity, and cardiovascular disease.

Niacin: Finally, I find that some patients, especially those with high triglycerides and low HDL (good cholesterol), benefit from niacin. Niacin is a form of vitamin B_3 found in many foods, including meat, fish, milk, eggs, green vegetables, beans, and fortified breads and cereals. It's required for the proper function of fats and sugars in the body and to maintain healthy cells, and because of its beneficial effects on clotting, it has been shown to help some people with heart disease. To benefit from niacin, you should take it only in the form of nicotinic acid (and as with all supplements, with your doctor's guidance). Nicotinic acid is the only form of niacin that seems to lower cholesterol, which is not contained in the "no flush" niacin preparations. (Some patients experience a flushing effect upon taking niacin, characterized by a burning, tingling, or itching sensation on the face, arms, and chest as well as redness in these areas and headaches.) It helps to start with small doses of niacin (50 mg), taking 81 to 325 mg of aspirin before each dose to help reduce the flushing reaction. This common reaction usually wanes as the body gets used to the medication.

It's important that you have a converstion with your doctor about exactly which drugs and supplements you should consider in your overall health strategy. I cannot outline every possible regimen, as each person is different depending on individual risks and health profiles. It's also critical that you and your doctor weigh the benefits against the potential risks involved with each drug or supplement. And, of course, you'll also have to consider how various drugs and supplements work with one another in the body when taken together.

Heart Prescriptions

I know the kind of worry and emotional struggle that heart patients from all walks of life carry, because I work with them every day. Some are in much more dire straits than others and must adhere to strict protocols to avoid serious injury to their heart, including a fatal heart attack.

My hope in this section is to quell that angst and sense of dread by giving you answers to the most common questions I get from my patients. I will start, however, by saying that regardless of my ideas here, you must make sure you feel that you're in confident, competent hands when it comes to the personal physician/clinician who is helping you address your condition on a continual basis. If you don't feel comfortable asking your doctor questions or following his or her recommendations, then find another doctor. Managing chronic heart disease is an ongoing process, so having a good relationship with your doctor is key. Now, let's get to those questions:

Which drugs (prescription or otherwise) are the most beneficial for me as a heart patient?

I give you two to consider, one that requires a prescription and another that you can get at your local drugstore or supermarket: (1) statins, for reasons I've already given, and (2) a daily baby aspirin (81 mg). Aspirin is safe if you have no bleeding issues, and especially beneficial for those with atherosclerosis. It's one of the oldest remedies known to humankind, and has far-reaching effects on the body as a whole that go beyond easing headaches and sore joints. Numerous studies have confirmed that the use of aspirin not only substantially reduces the risk of cardiovascular disease, but it can even ward off a host of ailments, cancer included, through its anti-inflammatory powers.

As a heart patient whose weight is normal, should I purposely gain weight?

Unless someone is truly underweight, it's not customary to tell anyone to gain weight or purposely add fat, irrespective of factors like fitness. Normal-weight patients with heart disease should first and foremost aim to increase their fitness level and maintain or increase muscle mass. This in turn can shift the body out of the hazard zone and may even encourage healthy weight gain through that increased muscle mass. Normal-weight individuals with heart disease likely have genetic forces working against them, for which being as fit as possible can help counter their DNA's codes at play.

How concerned should I be about my cholesterol levels?

Cholesterol levels can be very misleading, especially if you're just looking at the total number rather than the breakdown of LDL (bad) cholesterol versus HDL (good). Sometimes a person with a 160 mg/dl total cholesterol (who has a very low level of HDL) is worse off than an individual with a 260 mg/dl total cholesterol (who has a very high level of HDL). You need to know what your individual values are more so than the total, and then speak with your doctor about the best road to take in changing those numbers, if necessary. In general, you only need to be concerned if your bad cholesterol numbers are high, for which a cholesterol-lowering drug and attention to your diet are helpful. To get an accurate LDL reading, you must fast for, ideally, ten to twelve hours prior to your blood draw.

Which additional tests should I ask my doctor to perform on me?

In addition to the tests I've already outlined (see page 196), you may want to ask about getting a coronary artery calcium scan, if you're over the age of fifty and are at a higher risk for heart disease due to family history or some other indicator in your profile. This test is not generally covered by insurance but can now cost under one hundred dollars. It assesses calcified plaque in your arteries (a source of blockages that can lead to a heart attack), and if your levels are high, your doctor can suggest more aggressive treatments.

What can I be doing at home to monitor my condition?

You should be doing what I wish everyone would do, whether or not one has a heart condition. Monitor your blood pressure on a routine basis (daily, weekly, or monthly depending on your blood pressure risks and other conditions), using devices readily available at pharmacies for free or that you can buy cheap. Weigh in weekly or twice a month to prevent continued gaining, and track your exercise habits to be sure that you're getting enough physical activity.

Is there anything wrong with engaging in strenuous physical activity?

The short answer to that question is no, unless you're a middle-aged or older person with significant cardiovascular risk factors who has been sedentary for a considerable time. In that case, build up to being able to engage in strenuous physical activity by starting slow and adding more intensity as you can comfortably handle it. This is when getting specific guidance from your own doctor is helpful, since he or she can take into consideration your exact risk factors and condition. But don't let your diagnosis prevent you from engaging in exercise at all. You just need to tailor your program to your needs and limitations.

Should I avoid lifting heavy objects and heavy weight lifting?

If you have advanced heart disease, then I would advise that you avoid both lifting heavy objects and heavy weight lifting. Stick with lighter resistance training. As noted earlier, the same rule applies: Lift just enough weight to complete twelve to fifteen repetitions without considerably straining yourself. The risk involved with most heavy lifting, regardless of one's condition, is that it can cause substantial injury to muscles and bones, which then can force one to become sedentary.

Are there any supplements I should consider that specifically help heart patients?

You can benefit from the same ones I mentioned in the previous section: CoQ10 for muscular side effects from statins; omega-3s (commonly sold as fish oil supplements) if you do not eat much fatty fish; nicotinic acid under a clinician's guidance for certain lipid disorders (high cholesterol, high triglycerides, and low levels of HDL); and vitamin D if your levels are low.

Any particular dietary precautions to take?

Any healthy dietary protocol should focus on natural, wholesome foods that are low in bad fats (e.g., trans-fatty acids especially) and not super high in salt and processed sugars. For heart patients with severe heart failure or severe hypertension, I recommend avoiding both very high and very low amounts of salt in particular. Aim for a total of 3 to 6 grams per day. Although we tend to think about high salt intake as the bane, super-low-salt diets can also be detrimental, as shown in recent papers that I have published on this topic with Drs. James DiNicolantonio and James O'Keefe in *American Journal of Hypertension* and *The American Journal of Medicine*. And those with atherosclerosis should pay attention to their fat intake, avoiding the bad fats found in trans fats and processed foods. As discussed previously, when in doubt, think Mediterranean.

Is belly fat particularly bad for heart patients?

You'll recall from earlier chapters that studies examining people with heart disease have shown that those with more fat had a good prognosis especially *if they were fit*. And in a surprising twist of conventional wisdom, researchers at UCLA, led by Dr. Tamara Horwich, noted that among those with heart failure, higher belly fat was associated with a better prognosis. Results like this of course challenge the idea that all belly fat (and higher waist circumference) is categorically bad. But I don't think it forces us to immediately revise established ideas on the hazards of excess belly fat. It just means we have to conduct further research. In general, my heart studies indicate that people with diagnosed conditions are better off with high muscle and higher-than-normal fat (but not to the point of morbid obesity). But again, ideal pictures can turn ugly if there's a genetic predisposition to having high cholesterol, high blood pressure, or even high blood sugar. That person—despite his or her best efforts to maintain a healthy weight and fitness level—can still get heart disease.

A Fresh Start

Often, patients who are diagnosed with a chronic condition suddenly feel like they are beyond the help that old rules for healthy living offer, but to a large extent that's not the case; heart patients can often perform the same daily practices as their healthy counterparts so long as they use extra caution and modify where appropriate.

Although managing a health challenge on a daily basis can be overwhelming at times, it's also an opportunity. Once you get your wake-up call from a doctor, you have two options: Go on living as you always have and probably suffer the consequences with an early death; or "take heart." I find that the patients who do exceptionally well after a dreadful diagnosis are often the ones who use their condition as a constant reminder to engage in the kinds of lifestyle habits that have always translated to a long life. My wish is that you take my ideas to heart, gain a renewed sense of empowerment (despite any diagnosis now or in the future), and use the latest science to make better decisions and to be at peace with your body. After all, our health is up to each and every one of us in the end—nobody else.

Epilogue: Survival of the Fittest

Semmelweis reflex: any erroneous scientific theory or position irrationally held within the scientific community in the clear face of existing contrary evidence.

We take for granted today truisms that were once met with skepticism and had to be proven beyond a shadow of a doubt. The earth is round. What we eat and other lifestyle choices are related to our health. And the simple act of hand washing can prevent the spread of illness.

In the mid-nineteenth century, a Hungarian-born physician made a profound observation while working in the maternity clinic at the Vienna General Hospital, in Austria. Dr. Ignaz Semmelweis noticed that the number of cases of fatal puerperal fever, or childbed fever, was three times larger in the doctors' wards than in the midwives' wards. Puerperal fever was a harrowing disease, affecting women within the first three days after childbirth. It would progress rapidly, causing excruciating symptoms, including severe abdominal pain; fever; headache; a "cold fit" followed by extreme heat, perspiration, and thirst; nausea; vomiting; listlessness; constipation and diarrhea; and for some, total delirium. Many of the women stricken with this disease died. Although it had been recognized for centuries that women who'd just given birth were prone to fevers, the distinct name, "puerperal fever," first emerged in the historical record in the early eighteenth century. And no one at the time, not even Dr. Semmelweis, could have fathomed the real culprit: a bacterial infection of the upper genital tract. It would take science another several decades to discover germs and, much later, antibiotics to combat them.

But Dr. Semmelweis was onto something when he took note of the difference between mortality rates among new moms in two different wards. And like any curious, vigilant doctor, he started to test a few hypotheses based on what he was seeing, finding that the number of cases was drastically reduced if the doctors washed their hands carefully before treating a pregnant woman. He also noted that the risk of death was especially high among women who'd been treated by doctors who were previously in contact with corpses (explaining why the ward managed by midwives was less dangerous). So in 1847, Semmelweis introduced hand washing with chlorinated lime solutions for interns who had performed autopsies, and recorded stunning results. His "prescription" immediately reduced the incidence of fatal puerperal fever in the hospital from as high as 30 percent to about 1 to 2 percent.

Semmelweis made a bold conclusion for his era, suggesting that some unknown "cadaveric material" caused the fever and cleanliness mattered in prevention. That there was only one cause of the illness was an aggressive statement to make in his day. His observations went against the current scientific opinion of the time, which blamed diseases on an imbalance of certain fluids or "humors" in the body. The conventional wisdom was that diseases were attributed to many different and unrelated causes, and each case was as unique as each human. To dare to say that puerperal fever was the same in every woman and could be prevented by the same precaution—simple hand washing—was considered absurd in the eyes of his peers.

He lectured publicly about his ideas in 1850, but was met with hostility. Besides, his colleagues argued, even if his findings were correct, washing one's hands every time before treating a pregnant woman would be too much work. Behind the veil of this complaint, however, was certainly the doctors' hesitance to admit that they had caused so many deaths. Some doctors refused to believe that "a gentleman's hands" could transmit disease. It could not have helped, however, that Semmelweis failed to offer an acceptable scientific explanation for his findings.

Outraged by the indifference of his medical profession, Semmelweis began writing letters to prominent European obstetricians, at times accusing them of being irresponsible murderers for rejecting his extreme hypothesis. He spent fourteen years developing his theories and lobbying for their acceptance, but he continued to be ignored and ridiculed. Dismissed from the hospital and harassed by the Viennese medical community, he was eventually forced to move to Budapest.

Sadly, the book he wrote, *The Etiology, Concept, and Prophylaxis of Childbed Fever*, was published in 1861, just a few years before Louis Pasteur came onto the scene with his germ theory, which confirmed the fact that some diseases are caused by microorganisms. (Pasteur is credited with discovering the pathology of the puerperal fever, offering a theoretical explanation for Semmelweis's findings.) Semmelweis's book received poor reviews, and in 1865 he suffered a nervous breakdown. Committed to an insane asylum, he died there only two weeks later, at the age of forty-seven, possibly after being severely beaten by guards. And sadder still, only a few years after that, his hand-washing practice finally earned universal acceptance, thanks to Louis Pasteur and the work of Dr. Joseph Lister, who introduced the concept of using antiseptics to promote hygiene (and from whom Johnson & Johnson got the name of its popular oral antiseptic brand Listerine). In the 1870s, Lister was instrumental in developing practical applications of the germ theory of disease in medical settings. So while Semmelweis is considered a pioneer of antiseptic procedures, it was Lister whose name would go down in history.

We know a lot more today than any of these early physicians and scientists, but we still fall prey to what's now known as the Semmelweis reflex: a euphemism for the reflex-like tendency to reject new evidence or new knowledge because it contradicts established norms, beliefs, or paradigms. I can only imagine how frustrating it must have been for Dr. Semmelweis to experience such abandonment by his peers on a subject so essential and vital not only to his patients, but to society at-large. Perhaps I would have gone mad myself under those same circumstances.

What I love about science, which is why I've chosen to make a living practicing it within the medical sphere, is that there's never an absolute wrong or right answer. Solutions are often moving targets. That proverbial pendulum can swing to the far right for years or decades, only to turn back the other way for another indefinite time period upon the emergence of new knowledge. Perhaps this is why we tend to speak in terms of theories and hypotheses rather than categorical facts. We watch nature, record a set of observations, and then test out new ideas based on what we see. We open our data up for others to test out too, and address the scrutiny, criticism, or support. And perhaps this innate fickleness and uncertainty are why many of us in the scientific world also appreciate contradictions, for they give us the excuse to keep searching for better clues to understanding the complexities of life and, in particular, the human body.

When evidence to the contrary becomes apparent, we are forced to go back to the drawing board and sometimes discard an original theory, one that might have been widely accepted as indisputable fact. To think it was just a short while ago that we discovered the existence of infectious agents whose insults to the human body could be prevented or cured by vaccines, antibiotics, or in some cases, just good hygiene, is pretty remarkable.

Although we talk incessantly about the obesity epidemic dominating our health challenges in the twenty-first century, and many people view obesity in all of its forms as factually bad no matter what, my hope is that the narrative about fat shifts to one that's less weight-centric and more hinged on health defined by other measures. Because that's the most important message of all that the obesity paradox highlights: Health should not be measured by a number on the scale or the size of your jeans. It's much more complex than that. The person who measures health solely by weight is the equivalent of an early-eighteenth-century doctor measuring health by how "phlegmatic" or "sanguine" you are—two of the four symptoms associated with the four bodily humors, or fluids, whose forces were thought to control health. We know otherwise now.

We live in an exciting time in medicine, in which we not only have a much greater understanding of disease, aging processes, and the things we can do to prolong our lives, but we have the technology to more precisely customize health care based on one's unique physiology, genetics, value system, and conditions. We may be no better at achieving immortality than our prehistoric ancestors, but we're living much longer and better than any generation before us. Since the discovery of the germ theory, the average life span has soared from about forty-five to closer to eighty years. And if we do in fact see this number retreat, it won't be due to obesity per se. It'll be due to the real primary culprits of illness and disease: physical inactivity and poor nutrition that put the body on a dangerous path. Obesity can certainly be a coconspirator for the demise of some people, but more often than not it's an outcome of relentless insults the body has endured over time. So obesity, as it were, could really be viewed simply as the body's survivalist mechanisms in action to deal with those insults. And as we further document cases in which obesity appears to add to, rather than detract from, one's life span, we are forced to reconsider how to perceive of, define, and treat obesity.

A multitude of questions have yet to be answered, which will undoubtedly help us to push the boundaries of longevity further. Recent advances in and knowledge about things like the power of the human microbiome—the four or so pounds of bacteria that live in our intestinal tract—lead me to believe that the conversation about what it means to be "healthy" is just getting started. These foreign creatures that begin to colonize in our guts soon after our birth participate in digestion, metabolism, and overall health, including our weight and ability to prevent, manage, and cure disease. Researchers are currently looking at the possible role that some strains of intestinal bacteria have not only in obesity but also in inflammatory and gastrointestinal disorders, chronic pain, autism, depression, and autoimmune disorders such as rheumatoid arthritis, multiple sclerosis, and psoriasis. They are also looking into the role that these bacteria play in our emotions. Can a certain strain of bacteria in the digestive system make you generally happy and optimistic, or mean-spirited and pessimistic? We'll find out.

For the majority of our existence on the planet, being a little plump has been desirable, as reflected in the arts and literature. During the Renaissance, portrait paintings of royalty and upper-class individuals often showed off people's rounder figures. Peter

Paul Rubens was famous for his paintings of voluptuous, full-bodied women, from which we get the term *Rubenesque*. By the second half of the nineteenth century, however, excess fat was losing its aesthetic appeal; by the twentieth century, society was becoming more aware of its association with increased mortality. By then, fewer people were dying of infections and communicable diseases, thanks to rapid advances in medicine like those I just described. But also, fewer people were staying physically fit due to technologies that made them more sedentary in their jobs and day-to-day activities. This has become especially true in the past fifty years, as widespread availability of computers and automobiles, as well as general urbanization, has made immobility the norm. And now we succumb to chronic illnesses, like heart disease, stroke, cancer, and diabetes. Today the number of people dying from chronic afflictions is double that of people dying from infectious maladies, maternal and perinatal conditions, and nutritional deficiencies combined.

When I read a headline that says YOUNG ADULTS ONLY WALK FIVE MINUTES A DAY "BECAUSE OF TECHNOLOGY" and I think about the studies I've done that highlight the importance of fitness, I have to wonder if we've ignored the value of fitness to our detriment, especially where our children are concerned. It should be included as a vital sign. I also wonder if any decline in our life expectancy in the future could just be the result of too much sedentariness, regardless of what we eat and how fat we get. And much like having thinner, graying hair and more wrinkles as we grow older, maybe those extra pounds we carry in older age are a reflection of just that—age.

For decades we've pushed diets more so than exercise, and yet the number of chronic health challenges continues to climb. The fact that more than two-thirds of us are overweight or obese today and yet a breathtaking third of those people are healthy by virtually every measure besides weight means that we have to reconsider the implications of obesity and the role of fitness. Indeed, Charles Darwin had it right, around the same time Dr. Semmelweis was desperately trying to get his own ideas heard. A hundred and fifty years ago, Darwin foisted his grand theory about evolution on society. Although he wasn't talking in terms of physical fitness when he coined his famous "survival of the fittest" expression, he may as well have been referring to the survival benefits of being in decent physical shape. "Survival of the fittest" was originally a metaphor for "better designed for an immediate, local environment," and today I am convinced that physical fitness is exactly what we need to survive our immediate, local environment, which is rife with empty calories and other environmental hazards to health. Perhaps we'll find out in the future that those who can shoulder the weight of obesity but still survive longer than their thinner counterparts embody Darwin's theory of natural selection at work in the twenty-first century. Nature could be selecting for survival despite weight.

I know the value that being healthy brings to individuals' lives because I witness it daily in my own medical practice. I also see what sickness can do to people, no matter how much success they have experienced in life or how many, and how deeply, people love them. Without your health, you have nothing. But health isn't all about your physical body; it's also about your mind and psyche, which are also markedly improved by physical activity and exercise (as Dr. Richard Milani and I have published in numerous papers during the past two decades).

If we can loosen the reins on our obsession with weight and change the context of the conversation around our obesity crisis to our physical inactivity crisis, I think we can all move a little more and rest a little easier . . . and maybe live a lot longer and healthier.

Acknowledgments

Since I have been a practicing cardiologist and clinical investigator now going on three decades, there are many colleagues and patients who have contributed to the material in this book. Although this book reflects a lifetime of work in clinical care, teaching and clinical research, with numerous collaborators and thirty years of work related with obesity, my recognition and study of the specific area of the obesity paradox have been during the past decade. My main team of collaborators regarding the obesity paradox is with me here at the John Ochsner Heart and Vascular Institute in New Orleans, Louisiana, and particularly includes my very close friends and colleagues, Drs. Richard Milani and Hector Ventura, as well as a former colleague, Dr. Mandeep Mehra, who is now at Harvard. Several of my former cardiology fellows contributed to this important research, including Drs. Ahmed Osman, Dharmendrakumar "Dharm" Patel, and Surya Artham, and a current fellow who is doing very outstanding work in this area, Dr. Alban De Schutter. Many other residents and fellows have contributed to my research in preventive cardiology, including Drs. Arthur Menezes and Hassan Fares.

There are many outstanding scientists and clinicians from

around the United States and world who have worked on important projects with me regarding obesity and the paradox, most notably Dr. Steven Blair at the University of South Carolina. Steve was a leader for years of the Aerobic Center Longitudinal Study and data collected at the Cooper Institute years ago in Dallas, Texas, and he is one of the fathers of aerobic fitness and probably the best known in the world in the fitness versus fatness debate. Along with several of his colleagues at the University of South Carolina, including Dr. Xuemei ("Mei") Sui and D. C. Lee (who is now at Iowa State University), as well as Dr. Timothy Church, who has recently been the medical director of preventive medicine at the Pennington Biomedical Research Center (PBRC) in Baton Rouge, Louisiana, where I have also worked, and Dr. Paul McAuley from Winston-Salem State University, as well as many others, including colleagues Drs. Edward Archer, Gregory Hand, and Enrique Artero (now in Almeria, Spain), we have demonstrated the importance of physical activity and fitness in many groups of patients, including those with obesity and obesity-related disorders. Several other exercise scientists from PBRC, including Drs. Damon Swift (now at East Carolina University), Neil Johannsen (also at Louisiana State University), and Conrad Earnest (now at University of Bath, United Kingdom), and others from around the United States, including Dr. Ross Arena, who is a very close friend and colleague currently at University of Chicago in Illinois, and Dr. Daniel Forman, at Harvard, have also worked on several major projects together with me on exercise and fitness.

I would also like to thank Dr. Franz Messerli, now at St. Luke's Roosevelt Hospital Center, New York, who has been a mentor, colleague, and collaborator to me on the obesity paradox, as well as the many editors who have published some of my important original work, especially Drs. William Lanier of the *Mayo Clinic Proceedings*, Dr. Anthony DeMaria at *Journal of American College of Cardiology*, Dr. William C. Roberts of *The American Journal of Cardiology*, and Dr. Joseph Alpert at *The American Journal of Medicine*. Also, I thank my dear friend and close collaborator on many recent exercise projects, Dr. James O'Keefe, as well as Dr. James DiNicolantonio, who is doing excellent work in many aspects of preventive cardiology.

Finally, I thank some of my major other mentors, especially Drs. Bernard Gersh, Raymond Gibbons, Ray Squires, and Gerald Gau from my fellowship days at the Mayo Clinic.

Certainly, I give resounding thanks to my family, including my late father, Carl Joseph Lavie, and my mother, Gladys (Sue) Pfister Lavie, for supporting my siblings and me through all of our academic pursuits and for always promoting the importance of education and discipline to their children. I also have an amazing family for love and support, including my wife, Bonnie Bossetta Lavie, and our four adult children, three of whom (Jayson, Jennifer, and Scott) are also pursuing medical/physician careers, and Annie, who will provide some diversity for the family outside of the medical arena.

There are too many coworkers to thank, but I must thank my long-term medical assistant, Lucy Fraisse, long-term secretary, Patti Maggiore, current secretary, Jean Ann Landry, and medical editor, Patricia Taylor, as well as the many coworkers and colleagues who have provided support in so many special ways.

I also must thank the thousands of patients who have trusted me with their care for the past three decades. Thank you for your confidence in my care and for providing the stimulus for this book that hopefully provides information that will improve the understanding of obesity, physical activity, exercise, and physical fitness and their relative impacts in the prevention and treatment of cardiovascular diseases.

When my agent Bonnie Solow first contacted me about doing this book, I was excited about providing this information to a huge audience, but I was also concerned that all of my previous writing had been for academic audiences and not to the lay public. But Bonnie was such a tenacious and indefatigably passionate champion of my message that I had to say yes. This book, admittedly, is the brainchild of her "creative midwifery" and her commitment to representing works that have the power to offer a positive contribution to the world, potentially changing it. Thank you, Bonnie, for finding me and convincing me that my voice and message was unique, that this book was necessary, and that the world needed it. And indeed I am so grateful that I did finally agree to bring this science to the mainstream. Without her enthusiasm for cutting-edge health wisdom and an unwavering spirit for supporting messengers like me, I might never have bridged the connection between what I work with in my academic bubble and the public. Thank you, Bonnie, for also introducing me to my collaborator on this book, Kristin Loberg, who has spent countless hours skillfully transforming this medical information into a more easily digestible piece of writing.

Finally, a special thanks to Caroline Sutton and her team at the Hudson Street Press of the Penguin Group, for their confidence in our message and bringing this important information to the public.

—Carl "Chip" J. Lavie, MD

Notes

The following notes reflect specific facts and ideas mentioned in the text, and will open other doors for further research and inquiry. These are just a few of the many studies and data you can access today on the topics covered in the book. For additional reading recommendations and scientific papers that you might find helpful in learning more about some of the concepts revolving around the obesity paradox, I encourage you to visit my website, www.DrChipLavie.com, for ongoing and up-to-date resources.

Paradox: *Merriam-Webster Online*, s.v. "paradox," http://www.merriam-webster.com/dictionary/paradox.

Introduction

xi **more than two-thirds of us are overweight or obese:** For updated facts and figures about obesity, see the Centers for Disease Control and Prevention's website, accessed October 22, 2013, http://www.cdc.gov/obesity/data/adult.html.

xiv **the most obese state in the nation here in Louisiana:** For updated statistics on obesity in America, see the Centers for Disease Control and Prevention's website, accessed October 22, 2013. Figures can also be found at Trust for America's Health, a Washington, D.C.–based health policy organization that publishes an annual report titled *F as in Fat*; you can download the latest one at http://healthyamericans.org/report/108/.

xvii **and doctors endorsed cigarette smoking:** Stuart Elliott, "When Doctors, and Even Santa, Endorsed Tobacco," *New York Times*, October 6, 2008, http://www.nytimes.com/2008/10/07/business/media/07adco.html?_r=0.

Chapter 1

3 **"having eaten until fat":** *OED Online*, accessed October 22, 2013, http://www.oed.com.

3 ***The Oxford English Dictionary* documents:** *The Oxford English Dictionary*, 2nd ed. (Oxford University Press, 1989); *OED Online*, accessed October 22, 2013, http://www.oed.com/.

3 **In the sixth century BC:** S. Bhattacharya, "Sushrutha—Our Proud Heritage," *Indian Journal of Plastic Surgery* 42, no. 2 (2009): 223–5.

4 **"corpulence is not only a disease itself, but the harbinger of others":** C. Bain, "Public Health Aspects of Weight Control. Commentary: What's Past Is Prologue," *International Journal of Epidemiology* 35, no. 1 (2006): 16–7.

4 **Well over 70 percent of us:** http: "Overweight and Obesity," Centers for Disease Control and Prevention, last modified August 16, 2012, http://www.cdc.gov/obesity/data/adult.html.

4 **historically has been used to define obesity:** V. Hughes, "The Big Fat Truth," *Nature* 497, no. 7450 (2013): 428–30, doi: 10.1038/497428a.

5 **who is likely to die sooner rather than later:** Ibid.

5 **numbers started to go upward rapidly through the 1980s and 1990s:** P. T. von Hippel and R. W. Nahhas, "Extending the History of Child Obesity in the United States: The Fels Longitudinal Study, Birth Years 1930–1993," *Obesity* 21, no. 10 (2013): 2153–6, doi: 10.1002/oby.20395. See also: Steven D. Levitt and Stephen J. Dubner, "The History of Obesity Revisited," *Freakonomics* (blog), July 23, 2013, http://freakonomics.com/2013/07/23/the-history-of-obesity-revisited/.

6 **in the hundred years following the Civil War:** Ibid.

6 **seven times faster than the rate calculated over the previous hundred years:** Ibid.

6 **within a matter of just two generations:** Centers for Disease Control and Prevention, *Improving the Food Environment Through Nutrition Standards: A Guide for Government Procurement*, U.S. Department of Health and Human Services, Centers for Disease Control and Prevention, Natural Center for Chronic Disease Prevention and Health Promotion, Division for Heart Disease and Stroke Prevention (2011), http://www.cdc.gov/salt/pdfs/dhdsp_procurement

_guide.pdf. See also: H. A. Farah and J. C. Buzby, "U.S. Food Consumption Up 16 Percent since 1970," *Amber Waves* 3, no. 5 (2005): 5.

6 **accounted for the rest (188 calories):** Ibid.

6 **from those meals quadrupled:** Ibid.

6 **our obesity rate had gone up to 19.4 percent:** Centers for Disease Control and Prevention, "Early Release of Selected Estimates Based on Data from the 2004 National Health Interview Survey" (2005), http://www.cdc.gov/nchs/data/nhis/earlyrelease/200506_06.pdf.

6 **the World Health Organization (WHO) held its first meeting on the subject:** Hughes, "The Big Fat Truth."

7 **nearly 300 million women were obese:** National Center for Health Statistics, "Table 71: Overweight, Obesity, and Healthy Weight Among Persons 20 Years of Age and Over, by Selected Characteristics: United States, Selected Years 1960–62 through 2005–2008," *Health, United States, 2010* (Hyattsville, MD: U.S. Department of Health and Human Services, 2011), http://www.cdc.gov/nchs/data/hus/2010/071.pdf. See also: Centers for Disease Control and Prevention, "Overweight and Obesity"; World Health Organization, "Obesity and Overweight," last modified March 2013, http://www.who.int/mediacentre/factsheets/fs311/en/.

7 **Since the dawn of recorded history:** G. Eknoyan, "A History of Obesity, or How What Was Good Became Ugly and Then Bad," *Advances in Chronic Kidney Disease* 13, no. 4 (2006): 421–7. View the abstract at http://www.ackdjournal.org/article/S1548-5595%2806%2900106-6/abstract.

7 **to become so grand and reliable:** Ibid.

8 **the real culprit is energy expenditure:** E. Archer, R. P. Shook, D. M. Thomas, T. S. Church, P. T. Katzmarzyk, J. R. Hébert, K. L. McIver, G. A. Hand, C. J. Lavie, and S. N. Blair, "45-Year Trends in Women's Use of Time and Household Management Energy Expenditure," *PLoS One* 8, no. 2 (2013): e56620, doi: 10.1371/journal.pone.0056620.

8 **Many of my colleagues agree:** T. S. Church, D. M. Thomas, C. Tudor-Locke, P. T. Katzmarzyk, C. P. Earnest, R. Q. Rodarte, C. K. Martin, S. N. Blair, and C. Bouchard, "Trends Over 5 Decades in U.S. Occupation-Related Physical Activity and Their Associations with Obesity," *PLoS One* 6, no. 5 (2011): e19657, doi: 10.1371/journal.pone.0019657.

8 **household energy expenditure over the past forty-five years:** Archer et al., "45-Year Trends."

9 **"catastrophic levels":** Centers for Disease Control and Prevention, "Overweight and Obesity."

9 **are costing America 160 billion dollars per year:** J. Cawley and C. Meyerhoefer, "The Medical Care Costs of Obesity: An Instrumental Variables Approach," *Journal of Health Economics* 31, no. 1 (2012): 219–30, doi: 10.1016/j

.jhealeco.2011.10.003. See also: Greg Voakes, "Infographic: The Business of Obesity," BusinessInsider.com, August 15, 2012, http://www.businessinsider.com/infographic-the-business-of-obesity-2012-8.

9 **devour a total of 450 billion dollars annually:** Ibid.

9 **will have type 2 diabetes as an adult:** National Diabetes Information Clearinghouse, "Diabetes Overview," National Institute of Diabetes and Digestive and Kidney Diseases, National Institutes of Health, accessed October 21, 2013, http://diabetes.niddk.nih.gov/dm/pubs/overview/.

9 **According to the American Heart Association:** American Heart Association, American Stroke Association, *Understanding Childhood Obesity: 2011 Statistical Handbook* (Dallas: American Heart Association, 2011), http://www.heart.org/idc/groups/heart-public/@wcm/@fc/documents/downloadable/ucm_428180.pdf.

10 **and 30 percent or more are overweight:** Centers for Disease Control and Prevention, "Overweight and Obesity."

10 **86 percent of us will be overweight:** Y. Wang, M. A. Beydoun, L. Liang, B. Caballero, and S. K. Kumanyika, "Will All Americans Become Overweight or Obese? Estimating the Progression and Cost of the US Obesity Epidemic," *Obesity* 16, no. 10 (2008): 2323–30, doi: 10.1038/oby.2008.351.

10 **an extra $2,646 and the average woman an extra $4,879:** Associated Press, "Obesity's Yearly Cost: $4,879 for a Woman, $2,646 for a Man," *USA Today*, September 21, 2010, http://usatoday30.usatoday.com/yourlife/fitness/2010-09-21-obesity-costs_n.htm.

10 **2.5 percent less than their thinner counterparts:** Del Jones, "Obesity Can Mean Less Pay," *USA Today*, accessed October 21, 2013, http://usatoday30.usatoday.com/money/workplace/2002-09-04-overweight-pay-bias_x.htm.

10 **thinner peers doing the same work:** Roberta R. Friedman and Rebecca M. Puhl, *Rudd Report: Weight Bias: A Social Justice Issue* (New Haven, CT: Yale Rudd Center for Food Policy & Obesity, 2012), http://www.yaleruddcenter.org/resources/upload/docs/what/reports/Rudd_Policy_Brief_Weight_Bias.pdf.

10 **US has the highest rate of obesity:** Centers for Disease Control and Prevention, "Overweight and Obesity."

10 **a fascinating and destructive phenomenon:** "The History of Obesity Timeline," HistoWiki.com, last modified July 1, 2013, http://histowiki.com/2244/history/obesity-history-timeline/.

10 **uniquely human or human caused:** Ibid.

11 **starvation if they are inept predators:** Ibid.

11 **To my knowledge, the expression *obesity paradox*:** L. Gruberg, N. J. Weissman, R. Waksman, S. Fuchs, R. Deible, E. E. Pinnow, L. M. Ahmed, et al., "The Impact of Obesity on the Short-Term and Long-Term Outcomes After

Percutaneous Coronary Intervention: The Obesity Paradox?" *Journal of the American College of Cardiology* 39, no. 4 (2002): 578–84.

13 **than if they are normal weight:** Harriet Brown, "In 'Obesity Paradox,' Thinner May Mean Sicker," *New York Times*, September 17, 2012, http://www.nytimes.com/2012/09/18/health/research/more-data-suggests-fitness-matters-more-than-weight.html.

16 **"We are not so much born to run as born to walk":** J. H. O'Keefe and C. J. Lavie, "Run for Your Life . . . At a Comfortable Speed and Not Too Far," *Heart* 99, no. 8 (2013): 516–9, doi: 10.1136/heartjnl-2012-302886.

17 **Diabetes patients of normal weight:** V. Hainer and I. Aldhoon-Hainerová, "Obesity Paradox Does Exist," *Diabetes Care* 36, Suppl 2 (2013): S276–81, doi: 10.2337/dcS13-2023.

17 **Heavier dialysis patients:** T. B. Horwich and G. C. Fonarow, "Reverse Epidemiology Beyond Dialysis Patients: Chronic Heart Failure, Geriatrics, Rheumatoid Arthritis, COPD, and AIDS," *Seminars in Dialysis* 20, no. 6 (2007): 549–53. See also: K. Kalantar-Zadeh and J. D. Kopple, "Obesity Paradox in Patients on Maintenance Dialysis," *Contributions to Nephrology* 151 (2006): 57–69.

17 **suffering from heart disease:** C. J. Lavie, A. De Schutter, and R. V. Milani, "Is There an Obesity, Overweight, or Lean Paradox in Coronary Heart Disease? Getting to the 'Fat' of the Matter," *Heart* 99, no. 9 (2013): 596–8. doi: 10.1136/heartjnl-2012-303487.

17 **increased mortality in the elderly:** A. Oreopoulos, K. Kalantar-Zadeh, A. M. Sharmah, and G. C. Fonarow, "The Obesity Paradox in the Elderly: Potential Mechanisms and Clinical Implications," *Clinics in Geriatric Medicine* 25, no. 4 (2009): 643–59, viii, doi: 10.1016/j.cger.2009.07.005. See also: Horwich and Fonarow, "Reverse Epidemiology Beyond Dialysis Patients."

17 **an infection such as HIV live longer:** Horwich and Fonarow, "Reverse Epidemiology Beyond Dialysis Patients."

18 **But then Dr. Katherine Flegal's explosive paper:** K. M. Flegal, B. K. Kit, H. Orpana, and B. I. Graubard, "Association of All-Cause Mortality with Overweight and Obesity Using Standard Body Mass Index Categories: A Systematic Review and Meta-Analysis," *JAMA* 309, no. 1 (2013): 71–82, doi: 10.1001/jama.2012.113905.

19 **a mathematician, not a physician:** *Encyclopaedia Britannica*, s. v. "Adolphe Quetelet," accessed October 21, 2013, http://www.britannica.com/EBchecked/topic/487148/Adolphe-Quetelet.

20 **to enjoy a good, long life:** C. J. Lavie, A. De Schutter, D. Patel, S. M. Artham, and R. V. Milani, "Body Composition and Coronary Heart Disease Mortality—an Obesity or a Lean Paradox?" *Mayo Clinic Proceedings* 86, no. 9 (2011): 857–64, doi: 10.4065/mcp.2011.0092.

21 **In Abigail Saguy's illuminating book:** Abigail Saguy, *What's Wrong with Fat?* (New York: Oxford University Press, 2013).

22 **not a medical explanation:** Ibid., 16.

22 ***metabolically healthy obesity*:** J. Naukkarinen, S. Heinonen, A. Hakkarainen, J. Lundbom, K. Vuolteenaho, L. Saarinen, S. Hautaniemi, et al., "Characterising Metabolically Healthy Obesity in Weight-Discordant Monozygotic Twins," *Diabetologia*, October 8, 2013 (Epub ahead of print).

22 **fifty-six million overweight and obese individuals who have no such abnormalities:** Abigail Saguy, "If Obesity Is a Disease, Why Are So Many Obese People Healthy?" *Time*, June 24, 2013, http://ideas.time.com/2013/06/24/if-obesity-is-a-disease-why-are-so-many-obese-people-healthy/.

23 **The impact of good nutrition:** R. Estruch, E. Ros, and M. A. Martínez-González, "Mediterranean Diet for Primary Prevention of Cardiovascular Disease," *New England Journal of Medicine* 369, no. 7 (2013): 676–7, doi: 10.1056/NEJMc1306659.

23 **"the disease categorization may reinforce blame by raising the stakes":** Saguy, "If Obesity Is a Disease."

23 **neglect and child endangerment:** Ibid.

24 **The idea that BMI is a faulty measure:** K. M. Flegal and B. I. Graubard, "Estimates of Excess Deaths Associated with Body Mass Index and Other Anthropometric Variables," *American Journal of Clinical Nutrition* 89, no. 4 (2009): 1213–9, doi: 10.3945/ajcn.2008.26698.

Chapter 2

27 **more diseases than smoking, alcoholism, and poverty:** "The Health Risks of Obesity: Worse Than Smoking, Drinking, or Poverty," Rand Corporation, accessed October 21, 2013, http://www.rand.org/pubs/research_briefs/RB4549/index1.html.

27 **If current trends continue:** "Obesity Overtaking Smoking as America's Number One Killer," Medical News Today, March 9, 2004, http://www.medicalnewstoday.com/releases/6438.php.

29 **CT scans of 137 mummies:** R. C. Thompson, A. H. Allam, G. P. Lombardi, L. S. Wann, M. L. Sutherland, J. D. Sutherland, M. A. Soliman, et al., "Atherosclerosis Across 4000 Years of Human History: The Horus Study of Four Ancient Populations," *Lancet* 381, no. 9873 (2013): 1211–22, doi: 10.1016/S0140-6736(13)60598-X.

29 **Transplantation is harder:** National Heart, Lung, and Blood Institute, "What Is a Heart Transplant?" National Institutes of Health, January 3, 2012, http://www.nhlbi.nih.gov/health/health-topics/topics/ht/.

33 **obese people had a first heart attack 6.8 years earlier:** M. C. Madala, B. A. Franklin, A. Y. Chen, A. D. Berman, M. T. Roe, E. D. Peterson, E. M.

Ohman, et al., "Obesity and Age of First Non-ST-Segment Elevation Myocardial Infarction," *Journal of the American College of Cardiology* 52, no. 12 (2008): 979–85, doi: 10.1016/j.jacc.2008.04.067.

34 **factors like cholesterol and blood pressure:** J. Logue, H. M. Murray, P. Welsh, J. Shepherd, C. Packard, P. Macfarlane, S. Cobbe, I. Ford, and N. Sattar, "Obesity Is Associated with Fatal Coronary Heart Disease Independently of Traditional Risk Factors and Deprivation," *Heart* 97, no. 7 (2011): 564–8, doi: 10.1136/hrt.2010.211201.

37 **What Is "Metabolic Syndrome":** Mayo Clinic staff, "Metabolic Syndrome," Mayo Clinic, April 5, 2013, http://www.mayoclinic.com/health/metabolic%20syndrome/DS00522.

38 **Blood pressure is the force of blood:** National Heart, Lung, and Blood Institute, "What Is High Blood Pressure?" National Institutes of Health, August 2, 2012, http://www.nhlbi.nih.gov/health/health-topics/topics/hbp/.

38 **from coronary heart disease are over age sixty-five:** "Coronary Artery Disease—Coronary Heart Disease," American Heart Association, August 30, 2013, http://www.heart.org/HEARTORG/Conditions/More/MyHeartandStrokeNews/Coronary-Artery-Disease—Coronary-Heart-Disease_UCM_436416_Article.jsp. See also: T. Thom, N. Haase, W. Rosamond, V. J. Howard, J. Rumsfeld, T. Manolio, Z. Zheng, et al., "Heart Disease and Stroke Statistics—2006 Update: A Report from the American Heart Association Statistics Committee and Stroke Statistics Subcommittee," *Circulation* 113 (2006): e85–e151, http://circ.ahajournals.org/content/113/6/e85.full.

39 **Every forty-five seconds someone has a stroke:** Jose Vega, "Interesting Facts and Statistics About Stroke," About.com, September 22, 2008, http://stroke.about.com/od/strokestatistics/a/StrokeStats.htm.

39 **low levels of HDL (good) cholesterol:** V. Demarin, M. Lisak, S. Morović, and T. Cengić, "Low High-Density Lipoprotein Cholesterol as the Possible Risk Factor for Stroke," *Acta Clinica Croatica* 49, no. 4 (2010): 429–39.

39 **thirty million Americans who have abnormal blood fats:** A. Romero-Corral, V. K. Somers, J. Sierra-Johnson, Y. Korenfeld, S. Boarin, J. Korinek, M. D. Jensen, G. Parati, and F. Lopez-Jimenez, "Normal Weight Obesity: A Risk Factor for Cardiometabolic Dysregulation and Cardiovascular Mortality," *European Heart Journal* 31, no. 6 (2010): 737–46, doi: 10.1093/eurheartj/ehp487. See also: Martica Hearner, "Thin on the Outside, Obese on the Inside," MSN Healthy Living, accessed October 21, 2013, http://healthyliving.msn.com/diseases/heart-and-cardiovascular/thin-on-the-outside-obese-on-the-inside-1.

40 **increased risk for cancer:** For facts and data on cancer and obesity, see: "Obesity and Cancer Risk," National Cancer Institute at the National Institutes of Health, accessed October 21, 2013, http://www.cancer.gov/cancertopics/factsheet/Risk/obesity.

42 **Osteoarthritis is a common joint problem:** For facts and data on osteoarthritis, see: "Arthritis-Related Statistics," Centers for Disease Control and Prevention, last reviewed August 1, 2011, http://www.cdc.gov/arthritis/data_statistics/arthritis_related_stats.htm; and the Arthritis Foundation's website, www.arthritis.org.

43 **although many are not diagnosed:** See the National Sleep Foundation's comprehensive website for info and data on sleep apnea: www.sleepfoundation.org.

44 **sleep apnea causes hypertension:** "Severe Sleep Apnea Increases Risk of Uncontrolled High Blood Pressure," American Academy of Sleep Medicine, June 3, 2013, http://www.aasmnet.org/articles.aspx?id=3929.

44 **experience of sudden cardiac death:** A. S. Gami, E. J. Olson, W. K. Shen, R. S. Wright, K. V. Ballman, D. O. Hodge, R. M. Herges, D. E. Howard, and V. K. Somers, "Obstructive Sleep Apnea and the Risk of Sudden Cardiac Death: A Longitudinal Study of 10,701 Adults," *Journal of the American College of Cardiology* 62, no. 7 (2013): 610–6, doi: 10.1016/j.jacc.2013.04.080.

45 **while 8 percent had asthma:** J. Ma, L. Xiao, and S. B. Knowles, "Obesity, Insulin Resistance, and the Prevalence of Atopy and Asthma in US Adults," *Allergy* 65, no. 11 (2010): 1455–63, doi: 10.1111/j.1398-9995.2010.02402.x. See also: J. Ma and L. Xiao, "Association of General and Central Obesity and Atopic and Nonatopic Asthma in US Adults," *Journal of Asthma* 50, no. 4 (2013): 395–402, doi: 10.3109/02770903.2013.770014.

46 **Obesity and kidney disease go hand in hand:** C. Wickman and H. Kramer, "Obesity and Kidney Disease: Potential Mechanisms," *Seminars in Nephrology* 33, no. 1 (2013): 14–22, doi: 10.1016/j.semnephrol.2012.12.006. For general information on kidney disease, see the National Kidney Foundation's website, www.kidney.org.

48 **We've known for some time now:** For a general overview of inflammation, see: Afsar U. Ahmed, "An Overview of Inflammation: Mechanism and Consequences," *Frontiers in Biology* 6, no. 4 (2011): 274–81, http://link.springer.com/article/10.1007%2Fs11515-011-1123-9#page-1. See also: P. C. Calder, R. Albers, J. M. Antoine, S. Blum, R. Bourdet-Sicard, G. A. Ferns, G. Folkerts, et al., "Inflammatory Disease Processes and Interactions with Nutrition," *British Journal of Nutrition* 101, Suppl 1 (2009): S1–45, doi: 10.1017/S0007114509377867.

51 **lower levels of several inflammatory markers:** C. M. Phillips and I. J. Perry, "Does Inflammation Determine Metabolic Health Status in Obese and Nonobese Adults?" *Journal of Clinical Endocrinology & Metabolism* 98, no. 10 (2013): E1610–9, doi: 10.1210/jc.2013-2038.

51 **better functioning mitochondria:** Naukkarinen et al., "Characterising Metabolically Healthy Obesity."

Chapter 3

54 **especially when we consider location:** For an overview of body fat location, see: Kathleen Doheny, "The Truth About Fat: Everything You Need to Know About Fat, Including an Explanation of Which Is Worse—Belly Fat or Thigh Fat," WebMD, July 13, 2009, http://www.webmd.com/diet/features/the-truth-about-fat. See also: Gary Taubes, *Why We Get Fat: And What to Do About It* (New York: Random House, 2010).

56 **growth hormone can in fact prompt weight loss:** V. S. Bonert, J. D. Elashoff, P. Barnett, and S. Melmed, "Body Mass Index Determines Evoked Growth Hormone (GH) Responsiveness in Normal Healthy Male Subjects: Diagnostic Caveat for Adult GH Deficiency," *Journal of Clinical Endocrinology & Metabolism* 89, no. 7 (2004): 3397–401.

57 **From a very broad standpoint:** For a layman's overview of types of body fat, see: "Everything You Want to Know About Body Fat," Shape Up America!, accessed October 21, 2013, http://www.shapeup.org/bfl/basics1.html. See also: Harvard Health Publications, "Abdominal Fat and What to Do About It," Harvard University, December 2006, http://www.health.harvard.edu/newsweek/Abdominal-fat-and-what-to-do-about-it.htm.

58 **the concept of lipotoxicity:** Harvard Medical School Family Health Guide, "Abdominal Obesity and Your Health," Harvard University, accessed October 21, 2013, http://www.health.harvard.edu/fhg/updates/abdominal-obesity-and-your-health.shtml.

58 **According to a Mayo Clinic observational study:** "People of Normal Weight with Belly Fat at Highest Death Risk, Mayo Clinic Study Finds," Mayo Clinic, August 27, 2012, http://www.mayoclinic.org/news2012-rst/7052.html.

59 **stamp out cardiovascular risk factors:** K.N. Manolopoulos, F. Karpe, and K. N. Frayn, "Gluteofemoral Body Fat as a Determinant of Metabolic Health," *International Journal of Obesity* 34, no. 6 (2010): 949–59, doi: 10.1038/ijo.2009.286.

59 **triglycerides and raised blood pressure:** B. H. Goodpaster, S. Krishnaswami, T. B. Harris, A. Katsiaras, S. B. Kritchevsky, E. M. Simonsick, M. Nevitt, P. Holvoet, and A. B. Newman, "Obesity, Regional Body Fat Distribution, and the Metabolic Syndrome in Older Men and Women," *Archives of Internal Medicine* 165, no. 7 (2005): 777–83.

60 **increase one's risk for heart disease:** F. Benatti, M. Solis, G. Artioli, E. Montag, V. Painelli, F. Saito, L. Baptista, et al., "Liposuction Induces a Compensatory Increase of Visceral Fat Which Is Effectively Counteracted by Physical Activity: A Randomized Trial," *Journal of Clinical Endocrinology & Metabolism* 97, no. 7 (2012): 2388–95, doi: 10.1210/jc.2012-1012.

60 **In 1947, the French physician Jean Vague noticed:** For a well-cited and

thorough review of the history on our understanding of abdominal obesity and how we've come to define it, see: Harvard's School of Public Health, "Waist Size Matters," Harvard University, accessed October 21, 2013, http://www.hsph.harvard.edu/obesity-prevention-source/obesity-definition/abdominal-obesity/. See also: J. P. Després, "Health Consequences of Visceral Obesity," *Annals of Medicine* 33, no. 8 (2001): 534–41.

61 **measured their waist size and hip size:** C. Zhang, K. M. Rexrode, R. M. van Dam, T. Y. Li, and F. B. Hu, "Abdominal Obesity and the Risk of All-Cause, Cardiovascular, and Cancer Mortality: Sixteen Years of Follow-Up in US Women," *Circulation* 117, no. 13 (2009): 1658–67, doi: 10.1161/CIRCULATION AHA.107.739714.

61 **The Shanghai Women's Health Study:** X. Zhang, X. O. Shu, G. Yang, H. Li, H. Cai, Y. T. Gao, and W. Zheng, "Abdominal Adiposity and Mortality in Chinese Women," *Archives of Internal Medicine* 167, no. 9 (2007): 886–92.

61 **In 2007, for example:** L. de Koning, A. T. Merchant, J. Pogue, and S. S. Anand, "Waist Circumference and Waist-to-Hip Ratio as Predictors of Cardiovascular Events: Meta-Regression Analysis of Prospective Studies," *European Heart Journal* 28 (2007): 850–6.

62 **similarly strong predictors of type 2 diabetes:** G. Vazquez, S. Duval, D. R. Jacobs Jr., and K. Silventoinen, "Comparison of Body Mass Index, Waist Circumference, and Waist/Hip Ratio in Predicting Incident Diabetes: A Meta-Analysis," *Epidemiologic Reviews* 29 (2007): 115–28. See also: Q. Qiao and R. Nyamdorj, "Is the Association of Type II Diabetes with Waist Circumference or Waist-to-Hip Ratio Stronger Than That with Body Mass Index?" *European Journal of Clinical Nutrition* 64, no. 1 (2009): 30–4.

63 **individuals with a higher waist circumference actually fare *better*:** P. A. McAuley, E. G. Artero, X. Sui, D. C. Lee, T. S. Church, C. J. Lavie, J. N. Myers, V. España-Romero, and S. N. Blair, "The Obesity Paradox, Cardiorespiratory Fitness, and Coronary Heart Disease," *Mayo Clinic Proceedings* 87, no. 5 (2012): 443–51, doi: 10.1016/j.mayocp.2012.01.013. See also: A. L. Clark, G. C. Fonarow, and T. B. Horwich, "Waist Circumference, Body Mass Index, and Survival in Systolic Heart Failure: The Obesity Paradox Revisited," *Journal of Cardiac Failure* 17, no. 5 (2011): 374–80, doi: 10.1016/j.cardfail.2011.01.009.

63 **based on nearly ten thousand coronary artery disease:** McAuley et al., "The Obesity Paradox, Cardiorespiratory Fitness."

Chapter 4

65 **"sitting is the new smoking":** James A. Lavine, "What Are the Risks of Sitting Too Much?" Mayo Clinic, June 16, 2012, http://www.mayoclinic.com/

health/sitting/AN02082. See also: Chris Weller, "Is Sitting the New Smoking? A Workday of Inactivity Could Offset Any Benefits of Exercise," Medical Daily, July 29, 2013, http://www.medicaldaily.com/sitting-new-smoking-workday-inactivity-could-offset-any-benefits-exercise-248119. For a great summary of the findings, see: Selene Yeager, "Sitting Is the New Smoking—Even for Runners," *Runner's World*, July 20, 2013, http://www.runnersworld.com/health/sitting-is-the-new-smoking-even-for-runners. And for tips on how to reduce your time spent sitting at work, check out pp. 64–67 of Yeager's "The New Smoking?" *Runner's World*, August 2013, http://wellness.ua.edu/wp-content/uploads/2009/05/Is-Sitting-the-New-Smoking.pdf.

65 **The science showing this link between sitting time and total mortality:** A. V. Patel, L. Bernstein, A. Deka, H. S. Feigelson, P. T. Campbell, S. M. Gapstur, G. A. Colditz, and M. J. Thun, "Leisure Time Spent Sitting in Relation to Total Mortality in a Prospective Cohort of US Adults," *American Journal of Epidemiology* 172, no. 4 (2010): 419–29, doi: 10.1093/aje/kwq155.

66 **"spending 22 hours sitting on your rear end":** D. W. Dunstan, B. Howard, G. N. Healy, and N. Owen, "Too Much Sitting—A Health Hazard," *Diabetes Research and Clinical Practice* 97, no. 3 (2012): 368–76, doi: 10.1016/j.diabres.2012.05.020.

66 **spanning twelve years on more than seventeen thousand Canadians:** P. T. Katzmarzyk, T. S. Church, C. L. Craig, and C. Bouchard, "Sitting Time and Mortality from All Causes, Cardiovascular Disease, and Cancer," *Medicine and Science in Sports and Exercise* 41, no. 5 (2009): 998–1005, doi: 10.1249/MSS.0b013e3181930355.

66 **A key gene called *lipid phosphate phosphatase 1*, or LPP1, could be partly to blame:** T. W. Zderic and M. T. Hamilton, "Identification of Hemostatic Genes Expressed in Human and Rat Leg Muscles and a Novel Gene (LPP1/PAP2A) Suppressed During Prolonged Physical Inactivity (Sitting)," *Lipids in Health and Disease* 11 (2012): 137, doi:10.1186/1476-511X-11-137.

67 **researchers led by Marc Hamilton, PhD:** L. L. Craft, T. W. Zderic, S. M. Gapstur, E. H. Vaniterson, D. M. Thomas, J. Siddique, and M. T. Hamilton, "Evidence That Women Meeting Physical Activity Guidelines Do Not Sit Less: An Observational Inclinometry Study," *International Journal of Behavioral Nutrition and Physical Activity* 9 (2012): 122, doi: 10.1186/1479-5868-9-122.

68 **increased risk of both breast and colon cancers:** American Institute for Cancer Research, "How Sitting and Moving Link to Cancer Risk," *Cancer Research Update* 87, January 11, 2012, http://preventcancer.aicr.org/site/News2?id=21401.

68 **a 2013 survey of nearly thirty thousand women:** J. G. van Uffelen, Y. R. van Gellecum, N. W. Burton, G. Peeters, K. C. Heesch, and W. J. Brown, "Sitting-Time, Physical Activity, and Depressive Symptoms in Mid-Aged Women," *Amer-*

ican Journal of Preventive Medicine 45, no. 3 (2013): 276–81, doi: 10.1016/j.amepre.2013.04.009.

69 **"Endurance Running and the Evolution of *Homo*":** D. M. Bramble and D. E. Lieberman, "Endurance Running and the Evolution of Homo," *Nature* 432, no. 7015 (2004): 345–52.

71 **called *niacinamide* than unfit people:** G. D. Lewis, L. Farrell, M. J. Wood, M. Martinovic, Z. Arany, G. C. Rowe, A. Souza, et al., "Metabolic Signatures of Exercise in Human Plasma," *Science Translational Medicine* 2, no. 33 (2010): 33ra37, doi: 10.1126/scitranslmed.3001006.

72 **the brain's "growth hormone":** C. D. Wrann, J. P. White, J. Salogiannnis, D. Laznik-Bogoslavski, J. Wu, D. Ma, J. D. Lin, M. E. Greenberg, and B. M. Spiegelman, "Exercise Induces Hippocampal BDNF Through a PGC-1α/FNDC5 Pathway," *Cell Metabolism* 18, no. 5 (2013): 649–59, doi: 10.1016/j.cmet.2013.09.008.

73 **a third type of diabetes:** S. M. de la Monte and J. R. Wands, "Alzheimer's Disease Is Type 3 Diabetes—Evidence Reviewed," *Journal of Diabetes Science and Technology* 2, no. 6 (2008): 1101–13.

73 **twice as likely to develop Alzheimer's disease:** Joslin Diabetes Center, "Possible Mechanism for Link Between Diabetes and Alzheimer's Disease Discovered," *ScienceDaily*, February 17, 2004, http://www.sciencedaily.com/releases/2004/02/040217072709.htm.

73 **a team of Canadian and American researchers:** S. B. Wilkinson, S. M. Phillips, P. J. Atherton, R. Patel, K. E. Yarasheski, M. A. Tarnopolsky, and M. J. Rennie, "Differential Effects of Resistance and Endurance Exercise in the Fed State on Signalling Molecule Phosphorylation and Protein Synthesis in Human Muscle," *Journal of Physiology* 586, pt 15 (2008): 3701–17, doi: 10.1113/jphysiol.2008.153916.

74 **muscle cells by 40 to 50 percent:** D. A. Hood, G. Uguccioni, A. Vainshtein, and D. D'souza, "Mechanisms of Exercise-Induced Mitochondrial Biogenesis in Skeletal Muscle: Implications for Health and Disease," *Comprehensive Physiology* 1, no. 3 (2011): 1119–34, doi: 10.1002/cphy.c100074.

74 **a process called *methylation*:** For a lay review of the studies about gene methylation, see: Gretchen Reynolds, "How Exercise Changes Fat and Muscle Cells," *Well* (blog), *New York Times*, July 31, 2013, http://well.blogs.nytimes.com/2013/07/31/how-exercise-changes-fat-and-muscle-cells/?_r=0.

75 **important medical papers on this topic:** C. J. Lavie, R. J. Thomas, R. W. Squires, T. G. Allison, and R. V. Milani, "Exercise Training and Cardiac Rehabilitation in Primary and Secondary Prevention of Coronary Heart Disease," *Mayo Clinic Proceedings* 84, no. 4 (2009): 373–83, doi: 10.1016/S0025-6196(11)60548-X.

77 a **major cardiovascular event:** Lavie et al., "Body Composition and Coronary Heart Disease Mortality."

78 **according to the National Institutes of Health:** "Diabetes," MedlinePlus, last reviewed October 18, 2013, http://www.nlm.nih.gov/medlineplus/diabetes.html.

78 **odds of developing type 2 diabetes by 58 percent:** For a well-cited review on glucose metabolism, diabetes, and exercise, see: Harvard School of Public Health, "Simple Steps to Preventing Diabetes," Harvard University, accessed October 21, 2013, http://www.hsph.harvard.edu/nutritionsource/preventing-diabetes-full-story/.

79 **increase of 1 MET in cardiorespiratory fitness:** D. C. Lee, X. Sui, E. G. Artero, I. M. Lee, T. S. Church, P. A. McAuley, F. C. Stanford, H. W. Kohl 3rd, and S. N. Blair, "Long-Term Effects of Changes in Cardiorespiratory Fitness and Body Mass Index on All-Cause and Cardiovascular Disease Mortality in Men: The Aerobics Center Longitudinal Study," *Circulation* 124, no. 23 (2011): 2483–90, doi: 10.1161/CIRCULATIONAHA.111.038422.

79 **cancer than normal-weight people:** F. B. Ortega, D. C. Lee, P. T. Katzmarzyk, J. R. Ruiz, X. Sui, T. S. Church, and S. N. Blair, "The Intriguing Metabolically Healthy but Obese Phenotype: Cardiovascular Prognosis and Role of Fitness," *European Heart Journal* 34, no. 5 (2013): 389–97, doi: 10.1093/eurheartj/ehs174.

80 **when he said to *ScienceDaily*:** European Society of Cardiology, " 'Fitness and Fatness': Not All Obese People Have the Same Prognosis; Second Study Sheds Light on 'Obesity Paradox,' " *ScienceDaily*, September 4, 2012, http://www.sciencedaily.com/releases/2012/09/120904193052.htm.

81 **changes the relation of fatness to mortality:** McAuley et al., "The Obesity Paradox, Cardiorespiratory Fitness."

81 **impact of traditional risk factors on mortality:** Ortega et al., "The Intriguing Metabolically Healthy but Obese Phenotype."

83 **Researchers at UCLA:** P. Srikanthan and A. S. Karlamangla, "Relative Muscle Mass Is Inversely Associated with Insulin Resistance and Prediabetes. Findings from the Third National Health and Nutrition Examination Survey," *Journal of Clinical Endocrinology and Metabolism* 96, no. 9 (2011): 2898–903, doi: 10.1210/jc.2011-0435.

84 **Strength training can provide up to:** For a general overview of the benefits of strength training, see: "Why Strength Training?" Centers for Disease Control and Prevention, last modified February 24, 2011, http://www.cdc.gov/physicalactivity/growingstronger/why/.

84 **In most people, muscle strength peaks:** For a general review of muscle biology, see: Michael F. Holick, "Hard Bodies," in *The Vitamin D Solution: A*

3-Step Strategy to Cure Our Most Common Health Problems (New York: Plume, 2011).

85 **"The Underappreciated Role of Muscle in Health and Disease":** R. R. Wolfe, "The Underappreciated Role of Muscle in Health and Disease," *American Journal of Clinical Nutrition* 84, no. 3 (2006): 475–82.

87 **more than sixty days of fasting:** Ibid.

87 **February to the middle of July in 1942:** To read about the Warsaw Ghetto's experiments, see: Myron Winick, *Final Stamp: The Jewish Doctors in the Warsaw Ghetto* (Bloomington, IN: Author House, 2007).

87 **The extensive work by Ancel Keys:** L. M. Kalm and R. D. Semba, "They Starved So That Others Be Better Fed: Remembering Ancel Keys and the Minnesota Experiment," *Journal of Nutrition* 135, no. 6 (2005): 1347–52.

87 **to recover from a serious illness:** Wolfe, "The Underappreciated Role of Muscle."

88 **never walk again:** Ibid.

88 **It's increasingly understood that chronic diseases:** "Chronic Diseases and Health Promotion," Centers for Disease Control and Prevention, last modified August 13, 2012, http://www.cdc.gov/chronicdisease/overview/index.htm.

88 **how much muscle mass they lose:** Wolfe, "The Underappreciated Role of Muscle."

89 **a report from the Swedish Coronary Angiography and Angioplasty Registry:** O. Angerås, P. Albertsson, K. Karason, T. Råmunddal, G. Matejka, S. James, B. Lagerqvist, A. Rosengren, and E. Omerovic, "Evidence for Obesity Paradox in Patients with Acute Coronary Syndromes: A Report from the Swedish Coronary Angiography and Angioplasty Registry," *European Heart Journal* 34, no. 5 (2013): 345–53, doi: 10.1093/eurheartj/ehs217.

Chapter 5

93 **One of the most famous paradoxes:** J. Ferrières, "The French Paradox: Lessons for Other Countries," *Heart* 90, no. 1 (2004): 107–11.

95 **I first started to document this paradox:** C. J. Lavie, R. Milani, M. R. Mehra, H. O. Ventura, and F. H. Messerli, "Obesity, Weight Reduction and Survival in Heart Failure," *Journal of the American College of Cardiology* 39, no. 9 (2002): 1563; author reply 1563–4.

95 **There had been some recently published studies:** Horwich et al., "The Relationship Between Obesity and Mortality in Patients with Heart Failure," *Journal of the American College of Cardiology* 39, no. 3 (2001): 789–95.

97 **My initial study was finally published:** C. J. Lavie, A. F. Osman, R. V. Milani, and M. R. Mehra, "Body Composition and Prognosis in Chronic Sys-

tolic Heart Failure: The Obesity Paradox," *American Journal of Cardiology* 91, no. 7 (2003): 891–4.

97 **evaluated forty studies of more than 250,000 patients:** A. Romero-Corral, V. M. Montori, V. K. Somers, J. Korinek, R. J. Thomas, T. G. Allison, F. Mookadam, and F. Lopez-Jimenez, "Association of Bodyweight with Total Mortality and with Cardiovascular Events in Coronary Artery Disease: A Systematic Review of Cohort Studies," *Lancet* 368, no. 9536 (2006): 666–78.

97 **twenty-nine thousand patients from nine major heart failure studies:** A. Oreopoulos, R. Padwal, K. Kalantar-Zadeh, G. C. Fonarow, C. M. Norris, and F. A. McAlister, "Body Mass Index and Mortality in Heart Failure: A Meta-Analysis," *American Heart Journal* 156, no. 1 (2008): 13–22.

97 **a look at eleven thousand Canadians over more than a decade:** H. M. Orpana, J. M. Berthelot, M. S. Kaplan, D. H. Feeny, B. McFarland, and N. A. Ross, "BMI and Mortality: Results from a National Longitudinal Study of Canadian Adults," *Obesity* 18, no. 1 (2010): 214–8, doi: 10.1038/oby.2009.191.

97 **lowest chance of dying from any cause:** For a general overview of the obesity paradox and relevant studies, see: C. J. Lavie, A. De Schutter, D. A. Patel, T. S. Church, R. Arena, A. Romero-Corral, P. McAuley, H. O. Ventura, and R. V. Milani, "New Insights into the 'Obesity Paradox' and Cardiovascular Outcomes," *Journal of Glycomics and Lipidomics* 2 (2012): e106, doi:10.4172/2153-0637.1000e106.

99 **As early as 1982, obesity was linked to better survival:** P. Degoulet. M. Legrain, I. Réach, F. Aimé, C. Devriés, P. Rojas, and C. Jacobs, "Mortality Risk Factors in Patients Treated by Chronic Hemodialysis. Report of the Diaphane Collaborative Study," *Nephron* 31, no. 2 (1982): 103–10.

100 **In 1999, Dr. Erwin Fleischmann and colleagues:** E. Fleischmann, N. Teal, J. Dudley, W. May, J. D. Bower, and A. K. Salahudeen, "Influence of Excess Weight on Mortality and Hospital Stay in 1346 Hemodialysis Patients," *Kidney International* 55, no. 4 (1999): 1560–7.

100 **extra calories and reserves that lower their risk of death:** K. Kalantar-Zadeh and J. D. Kopple, "Obesity Paradox in Patients on Maintenance Dialysis," *Contributions to Nephrology* 151 (2006): 57–69. See also: T. Vashistha, R. Mehrotra, J. Park, E. Streja, R. Dukkipati, A. R. Nissenson, J. Z. Ma, C. P. Kovesdy, and K. Kalantar-Zadeh, "Effect of Age and Dialysis Vintage on Obesity Paradox in Long-Term Hemodialysis Patients," *American Journal of Kidney Diseases*, October 9, 2013, pii: S0272-6386(13)01113-X, doi: 10.1053/j.ajkd.2013.07.021 (Epub ahead of print).

100 **In 2012, diabetes researcher Dr. Mercedes Carnethon:** M. R. Carnethon, P. J. De Chavez, M. L. Biggs, C. E. Lewis, J. S. Pankow, A. G. Bertoni, S. H. Golden, et al., "Association of Weight Status with Mortality in Adults with

Incident Diabetes," *JAMA* 308, no. 6 (2012): 581–90, doi: 10.1001/jama.2012.9282.

101 **researchers at Tel Aviv University:** E. Osher and N. Stern, "Obesity in Elderly Subjects: In Sheep's Clothing Perhaps, but Still a Wolf!" *Diabetes Care* 32, Suppl 2 (2009): S398–402, doi: 10.2337/dc09-S347. See also: A. Oreopoulos, K. Kalantar-Zadeh, A. M. Sharma, and G. C. Fonarow, "The Obesity Paradox in the Elderly: Potential Mechanisms and Clinical Implications," *Clinics in Geriatric Medicine* 25, no. 4 (2009): 643–59, viii, doi: 10.1016/j.cger.2009.07.005.

102 **In 2005, researchers at the University of Texas Health Science Center at San Antonio:** A. Escalante, R. W. Haas, and I. del Rincón, "Paradoxical Effect of Body Mass Index on Survival in Rheumatoid Arthritis: Role of Comorbidity and Systemic Inflammation," *Archives of Internal Medicine* 165, no. 14 (2005): 1624–9.

103 **The Seven Countries Study:** A. Keys, A. Menotti, C. Aravanis, H. Blackburn, B. S. Djordevic, R. Buzina, A. S. Dontas, et al., "The Seven Countries Study: 2,289 Deaths in 15 Years," *Preventive Medicine* 13, no. 2 (1984): 141–54.

104 **Although body fat releases a plethora of bad molecules:** B. E. Wisse, "The Inflammatory Syndrome: The Role of Adipose Tissue Cytokines in Metabolic Disorders Linked to Obesity," *Journal of the American Society of Nephrology* 15, no. 11 (2004): 2792–800.

105 **reducing some of the deleterious effects:** A. L. Groeger, C. Cipollina, M. P. Cole, S. R. Woodcock, G. Bonacci, T. K. Rudolph, V. Rudolph, B. A. Freeman, and F. J. Schopfer, "Cyclooxygenase-2 Generates Anti-Inflammatory Mediators from Omega-3 Fatty Acids," *Nature Chemical Biology* 6, no. 6 (2010): 433–41, doi: 10.1038/nchembio.367. See also: P. C. Calder, "N-3 Polyunsaturated Fatty Acids, Inflammation, and Inflammatory Diseases," *American Journal of Clinical Nutrition* 83, Suppl 6 (2006): 1505S–519S.

105 **Oxford University researchers:** K. N. Manolopoulos, F. Karpe, and K. N. Frayn, "Gluteofemoral Body Fat as a Determinant of Metabolic Health," *International Journal of Obesity* 34, no. 6 (2010): 949–59, doi: 10.1038/ijo.2009.286.

106 **research like the Framingham Heart Study:** For all you want to know about this comprehensive study, go to http://www.framinghamheartstudy.org/.

107 **preserving mitochondrial health:** A. Safdar, J. P. Little, A. J. Stokl, B. P. Hettinga, M. Akhtar, and M. A. Tarnopolsky, "Exercise Increases Mitochondrial PGC-1alpha Content and Promotes Nuclear-Mitochondrial Cross-Talk to Coordinate Mitochondrial Biogenesis," *Journal of Biological Chemistry* 286, no. 12 (2011): 10605–17, doi: 10.1074/jbc.M110.211466.

107 **The (Unfortunate) Power of Bias:** For a compelling perspective on the topic of biases in the obesity conversation, see: Saguy, *What's Wrong with Fat?*

108 **In 1981 Dr. Neil Ruderman:** N. B. Ruderman, S. H. Schneider, and

P. Berchtold, "The 'Metabolically-Obese,' Normal-Weight Individual," *American Journal of Clinical Nutrition* 34, no. 8 (1981): 1617–21.

109 **(fitting for diet season):** Flegal et al., "Association of All-Cause Mortality."

109 **But it really began in 2005:** K. M. Flegal, B. I. Graubard, D. F. Williamson, and M. H. Gail, "Excess Deaths Associated with Underweight, Overweight, and Obesity," *JAMA* 293, no. 15 (2005): 1861–7.

109 **encompassing a staggering 2.9 million people:** Flegal et al., "Association of All-Cause Mortality."

110 **"Science Is Truth Found Out":** Ed Ruscha, the author of this quote, which is part of his artwork in our National Gallery of Art, is famously noncommittal (he declines to offer any explanations about the meaning—if any—behind his art). But I think his statement speaks volumes without explanation.

110 **compelling astute observers to take note:** For an excellent and thorough review of the obesity record and history of defining it, see: Hughes, "The Big Fat Truth."

111 **In a frequently cited JAMA paper published in 1987:** J. E. Manson, M. J. Stampfer, C. H. Hennekens, and W. Willett, "Body Weight and Longevity: A Reassessment," *JAMA* 257, no. 3 (1987): 353–8.

112 **Analyzing data from 1.46 million people:** A. Berrington de Gonzalez, P. Hartge, J. R. Cerhan, A. J. Flint, L. Hannan, R. J. MacInnis, S. C. Moore, et al., "Body-Mass Index and Mortality Among 1.46 Million White Adults," *New England Journal of Medicine* 363, no. 23 (2010): 2211–9, doi: 10.1056/NEJ-Moa1000367.

113 **"a toxic, inflammatory envelope":** Hughes, "The Big Fat Truth."

116 **"make measurable what cannot be measured":** R. S. Ahima and M. A. Lazar, "Physiology. The Health Risk of Obesity—Better Metrics Imperative," *Science* 341, no. 6148 (2013): 856–8, doi: 10.1126/science.1241244.

116 **within six years of surviving a heart attack:** T. Thom et al., "Heart Disease and Stroke Statistics."

Chapter 6

121 **the humble fat cell:** For a general overview of what fats do in the body, see: Stephanie Dutchen, "What Do Fats Do in the Body?" National Institute of General Medical Sciences, December 15, 2010, http://publications.nigms.nih.gov/insidelifescience/fats_do.html. Also see: Shape Up America!, "Everything You Want to Know."

122 **the most efficient source of food energy:** For an overview of fat's importance, as well as the different types, see: Harvard School of Public Health, "Fats and Cholesterol: Out with the Bad, in with the Good," Harvard University, accessed October 22, 2013, http://www.hsph.harvard.edu/nutritionsource/fats

-full-story/. Also see: *Encyclopaedia Britannica*, s. v. "fat," accessed October 22, 2013, http://www.britannica.com/EBchecked/topic/202365/fat.

123 **understand some basics:** Ibid.

125 **and attention deficit hyperactivity disorder (ADHD):** For a complete and well-cited synthesis of studies about fats and brain health, see: David Perlmutter and Kristin Loberg, Grain Brain: The Surprising Truth About Wheat, Carbs, and Sugar—Your Brain's Silent Killers (New York: Little, Brown, 2013), chapters 3 and 6.

128 **helpful if you have type 2 diabetes:** For an overview of dietary fats and how they affect the body, see: Mayo Clinic Staff, "Dietary Fats: Know Which Types to Choose," Mayo Clinic, February 15, 2011, http://www.mayoclinic.com/health/fat/NU00262.

129 **high total cholesterol levels, at least late in life, are not always associated with increased longevity:** D. Jacobs, H. Blackburn, M. Higgins, D. Reed, H. Iso, G. McMillan, J. Neaton, et al., "Report of the Conference on Low Blood Cholesterol: Mortality Associations," *Circulation* 86, no. 3 (1992): 1046–60. Also see: Perlmutter and Loberg, *Grain Brain*.

130 **Of all the functions and purposes of fat in the body:** To read a comprehensive exploration of body fat, check out Gary Taubes's works, including *Good Calories, Bad Calories: Fats, Carbs, and the Controversial Science of Diet and Health* (New York: Anchor, 2008) and *Why We Get Fat*.

134 **featuring autoimmune diseases or immune deficiencies:** D. A. Papanicolaou, R. L. Wilder, S. C. Manolagas, and G. P. Chrousos, "The Pathophysiologic Roles of Interleukin-6 in Human Disease," *Annals of Internal Medicine* 128, no. 2 (1998): 127–37.

137 **hormone-like substances that may be directly contributing to weight gain:** For a summary of obesogens, see: W. Holtcamp, "Obesogens: An Environmental Link to Obesity," *Environmental Health Perspectives* 120, no. 2 (2012): a62–8, doi: 10.1289/ehp.120-a62.

138 **damage normal function of metabolic hormones:** To access studies about environmental toxins that can affect metabolism and weight, see the Environmental Working Group's website, www.ewg.org.

Chapter 7

142 **Fat's Active Life: The Skinny on Fat Metabolism:** For a comprehensive exploration of fat metabolism in layman's terms, see: Taubes, *Why We Get Fat*. For an ongoing, interesting debate about fat metabolism, see: *The Spark of Reason* (blog), accessed October 22, 2013, sparkofreason.blogspot.com. For more about the different types of fat, including fat that burns calories, I encourage you to check out the research of Harvard's Dr. Bruce Spiegelman. The following

article will get you started: Elizabeth Gudrais, "Oxidative Oxymoron: The Fit Fat," *Harvard Magazine*, January–February 2009, http://harvardmagazine.com/2009/01/the-fit-fat.

145 **It also orchestrates our inflammatory response:** For a summary of leptin's chief responsibilities, see: Nora T. Gedgaudas, *Primal Body, Primal Mind: Beyond the Paleo Diet for Total Health and a Longer Life* (Rochester, VT: Healing Arts Press, 2011), chapter 14.

145 **A now seminal study published in 2004:** S. Taheri, L. Lin, D. Austin, T. Young, and E. Mignot, "Short Sleep Duration Is Associated with Reduced Leptin, Elevated Ghrelin, and Increased Body Mass Index," *PLoS Medicine* 1, no. 3 (2004): e62.

149 **"has to settle for expending less energy":** Taubes, *Why We Get Fat*.

Chapter 8

151 **the following headlines hit:** The following links relate to these headlines: Maggie Fox, "Heavy Burden: Obesity May Be Even Deadlier Than Thought," NBC News, August 15, 2013, http://www.nbcnews.com/health/heavy-burden-obesity-may-be-even-deadlier-thought-6C10930019; Columbia University's Mailman School of Public Health, "Obesity Kills More Americans Than Previously Thought: One in Five Americans, Black and White, Die from Obesity," *ScienceDaily*, August 15, 2013, http://www.sciencedaily.com/releases/2013/08/130815172339.htm; Rachel Auerbach, "Obesity Kills More Americans Than We Thought," *The Chart* (blog), CNN.com, August 15, 2013, http://thechart.blogs.cnn.com/2013/08/15/obesity-kills-more-americans-than-we-thought/; David L. Katz, "No More Denial: Obesity Kills," Everyday Health, August 15, 2013, http://www.everydayhealth.com/columns/health-answers/no-more-denial-obesity-kills-david-katz-md/; and Dennis Thompson, "Obesity's Death Toll May Be Higher Than Thought," WebMD, August 15, 2013, http://www.webmd.com/diet/news/20130815/obesitys-death-toll-may-be-much-higher-than-thought.

151 **The news was based on a study:** R. K. Masters, E. N. Reither, D. A. Powers, Y. C. Yang, A. E. Burger, and B. G. Link, "The Impact of Obesity on US Mortality Levels: The Importance of Age and Cohort Factors in Population Estimates," *American Journal of Public Health* 103, no. 10 (2013): 1895–901, doi: 10.2105/AJPH.2013.301379.

152 **cut that number down to 112,000:** For a Q&A statement from the CDC, see: "Frequently Asked Questions About Calculating Obesity-Related Risk," Centers for Disease Control and Prevention, accessed October 22, 2013, http://www.cdc.gov/PDF/Frequently_Asked_Questions_About_Calculating_Obesity-Related_risk.pdf.

154 **top ten causes of death in this country:** "Leading Causes of Death," Centers for Disease Control and Prevention, last modified January 11, 2012, http://www.cdc.gov/nchs/fastats/lcod.htm.

156 **a quarter of those who bring in at least fifty thousand dollars per year:** To view a collection of statistics and facts about fat, see: Trust for America's Health, *F as in Fat*. These numbers were found in the 2012 version.

156 **I'll let you be the judge:** To view any of these headlines, plus more, google "Lavie, coffee, mortality."

157 **Our decades-long study:** J. Liu, X. Sui, C. J. Lavie, J. R. Hebert, C. P. Earnest, J. Zhang, and S. N. Blair, "Association of Coffee Consumption with All-Cause and Cardiovascular Disease Mortality," *Mayo Clinic Proceedings* 88, no. 10 (2013): 1066–74, doi: 10.1016/j.mayocp.2013.06.020.

158 **"The focus should be on moderation":** Ryan Jaslow, "Four Cups of Coffee a Day May Raise Early Death Risk in Younger Adults," CBS News, August 15, 2013, http://www.cbsnews.com/8301-204_162-57598689/.

158 **We used to think that ulcers were directly caused by stress:** For a summary of the story of *H. pylori*, see: Todd Neale, "25 Years in Peptic Ulcers: From Chronic to Curable," ABC News, December 25, 2009, http://abcnews.go.com/Health/PainManagement/25-years-peptic-ulcers-chronic-curable/story?id=9417292.

159 **"Correlation does not imply causation":** For a general definition of the phrase, see: "Correlation Does Not Imply Causation," Princeton University, accessed October 22, 2013, http://www.princeton.edu/~achaney/tmve/wiki100k/docs/Correlation_does_not_imply_causation.html.

160 **As with any logical fallacy:** Vladica M. Velićković, "Opinion: Statistical Misconceptions," *The Scientist*, July 31, 2013, http://www.the-scientist.com/?articles.view/articleNo/36781/title/Opinion—Statistical-Misconceptions/.

160 **In 2013, award-winning economist Emily Oster:** Emily Oster, *Expecting Better: Why the Conventional Pregnancy Wisdom Is Wrong—and What You Really Need to Know* (New York: Hudson Street Press, 2013).

160 **SPONGEBOB THREAT TO CHILDREN, RESEARCHERS ARGUE:** Roni Caryn Rabin, "Is SpongeBob SquarePants Bad for Children?" *Well* (blog), *New York Times*, September 12, 2011, http://well.blogs.nytimes.com/2011/09/12/is-spongebob-squarepants-bad-for-children/?_r=0.

162 **obesity and the onset of type 2 diabetes:** For a review of relevant studies, see the Obesity Society's website, http://www.obesity.org.

163 **a glitch in the immune system related to a certain protein:** E. Stolarczyk, C. T. Vong, E. Perucha, I. Jackson, M. A. Cawthorne, E. T. Wargent, N. Powell, J. B. Canavan, G. M. Lord, and J. K. Howard, "Improved Insulin Sensitivity Despite Increased Visceral Adiposity in Mice Deficient for the Immune

Cell Transcription Factor T-bet," *Cell Metabolism* 17, no. 4 (2013): 520–33, doi: 10.1016/j.cmet.2013.02.019.

164 **she's thin on the outside and fat on the inside:** Associated Press, "Thin People Can Be Fat on the Inside," NBC News, May 11, 2007, http://www.nbcnews.com/id/18594089/#.UmghlySQw-8.

164 **will be diabetic by the year 2050:** You can search for several statistics and projections about diabetes at the CDC's website, http://www.cdc.gov.

165 **"The Thin Man's Diabetes":** Jeff O'Connell, "The Thin Man's Diabetes," *Men's Health*, April 16, 2008, http://www.menshealth.com/health/deadly-truth-about-diabetes.

171 **According to researchers at Monash University:** M. L. Borg, Z. B. Andrews, E. J. Duh, R. Zechner, P. J. Meikle, and M. J. Watt, "Pigment Epithelium-Derived Factor Regulates Lipid Metabolism via Adipose Triglyceride Lipase," *Diabetes* 60, no. 5 (2011): 1458–66, doi: 10.2337/db10-0845.

171 **Research from the Joslin Diabetes Center:** X. Ma, J. H. Warram, V. Trischitta, and A. Doria, "Genetic Variants at the Resistin Locus and Risk of Type 2 Diabetes in Caucasians," *Journal of Clinical Endocrinology and Metabolism* 87, no. 9 (2002): 4407–10.

172 **endoplasmic reticulum—is stressed by obesity:** G. S. Hotamisligil, "Endoplasmic Reticulum Stress and the Inflammatory Basis of Metabolic Disease," *Cell* 140, no. 6 (2010): 900–17, doi: 10.1016/j.cell.2010.02.034.

Chapter 9

175 **marathon runners develop some degree of water intoxication:** B. D. Levine and P. D. Thompson, "Marathon Maladies," *New England Journal of Medicine* 352, no. 15 (2005): 1516–8.

176 **we can turn to a longitudinal study:** C. P. Wen, J. P. Wai, M. K. Tsai, Y. C. Yang, T. Y. Cheng, M. C. Lee, H. T. Chan, C. K. Tsao, S. P. Tsai, and X. Wu, et al., "Minimum Amount of Physical Activity for Reduced Mortality and Extended Life Expectancy: A Prospective Cohort Study," *Lancet.* 2011 Oct 1; 378(, no. 9798 (2011): 1244–53, doi: 10.1016/S0140-6736(11)60749-6. Epub 2011 Aug 16.

179 **We've in fact documented these subtle signs of heart damage:** H. R. Patil, J. H. O'Keefe, C. J. Lavie, A. Magalski, R. A. Vogel, and P. A. McCullough, "Cardiovascular Damage Resulting from Chronic Excessive Endurance Exercise," *Missouri Medicine* 109, no. 4 (2012): 312–21.

180 **data on runners and life expectancy:** O'Keefe and Lavie, "Run for Your Life."

181 **low-distance runners fared much better than the very high-distance runners:** P. Schnohr, J. L. Marott, P. Lange, and G. B. Jensen, "Longevity in

Male and Female Joggers: The Copenhagen City Heart Study," *American Journal of Epidemiology* 177, no. 7 (2013): 683–9, doi: 10.1093/aje/kws301.

Chapter 10

195 **"200-pound anorexics":** Melissa Dahl, "The 200-Pound Anorexic: Obese Teens at Risk for Disorder, but It's Often Unrecognized," *Today*, September 22, 2013, http://www.today.com/health/200-pound-anorexic-obese-teens-risk-disorder-its-often-unrecognized-4B11216388.

195 **have an eating disorder sometime in their life:** Ibid.

199 **by the quality and amount of sleep we get a night:** For a comprehensive discussion of sleep and updates on studies, see the National Sleep Foundation's website, www.sleepfoundation.org, and Dr. Michael Breus's website, www.drbreus.com. Dr. Breus is a leading sleep doctor in America.

201 **"maintain, don't gain":** G. G. Bennett, P. Foley, E. Levine, J. Whiteley, S. Askew, D. M. Steinberg, B. Batch, et al., "Behavioral Treatment for Weight Gain Prevention Among Black Women in Primary Care Practice: A Randomized Clinical Trial," *JAMA Internal Medicine* 173, no. 19 (2013): 1770–7, doi: 10.1001/jamainternmed.2013.9263.

203 **than those on a typical low-fat diet:** Estruch, Ros, and Martínez-González, "Mediterranean Diet for Primary Prevention."

204 **Myths about weight loss and even obesity:** K. Casazza, R. Pate, and D. B. Allison, "Myths, Presumptions, and Facts about Obesity," *New England Journal of Medicine* 368, no. 23 (2013): 2236–7, doi: 10.1056/NEJMsa1208051. For a good summary of the study's main takeaways, see: Gina Kolata, "Myths of Weight Loss Are Plentiful, Researcher Says," *Well* (blog), *New York Times*, January 30, 2013, http://well.blogs.nytimes.com/2013/01/30/myths-of-weight-loss-are-plentiful-researcher-says/.

208 **Unproven Ideas:** Ibid.

209 **Know the Facts about Weight Loss:** Ibid.

211 **stressed the importance of cardiorespiratory fitness:** L. A. Kaminsky, R. Arena, T. M. Beckie, P. H. Brubaker, T. S. Church, D. E. Forman, B. A. Franklin, et al., "The Importance of Cardiorespiratory Fitness in the United States: The Need for a National Registry: A Policy Statement from the American Heart Association," *Circulation* 127, no. 5 (2013): 652–62, doi: 10.1161/CIR.0b013e31827ee100.

215 **potentially increasing your risk for certain cancers:** E. A. Klein, I. M. Thompson Jr., C. M. Tangen, J. J. Crowley, M. S. Lucia, P. J. Goodman, L. M. Minasian, et al., "Vitamin E and the Risk of Prostate Cancer: The Selenium and Vitamin E Cancer Prevention Trial (SELECT)," *JAMA* 306, no. 14 (2011): 1549–56, doi: 10.1001/jama.2011.1437.

216 **in many of these people, high cholesterol is not a problem:** Heart

Protection Study Collaborative Group, "MRC/BHF Heart Protection Study of Cholesterol Lowering with Simvastatin in 20,536 High-Risk Individuals: A Randomised Placebo-Controlled Trial," *Lancet* 360, no. 9326 (2002): 7–22.

216 **strokes or heart attacks by about 40 percent:** K. L. Furie, S. E. Kasner, R. J. Adams, G. W. Albers, R. L. Bush, S. C. Fagan, J. L. Halperin, et al., "Guidelines for the Prevention of Stroke in Patients with Stroke or Transient Ischemic Attack: A Guideline for Healthcare Professionals from the American Heart Association/American Stroke Association," *Stroke* 42, no. 1 (2011): 227–76, doi: 10.1161/STR.0b013e3181f7d043.

216 **lowered risk of death from cancer among those who took statins:** S. F. Nielsen, B. G. Nordestgaard, and S. E. Bojesen, "Statin Use and Reduced Cancer-Related Mortality," *New England Journal of Medicine* 367, no. 19 (2012): 1792–802, doi: 10.1056/NEJMoa1201735.

216 **athletes over age fifty who were on statins and taking 200 mg of CoQ10 daily:** R. E. Deichmann, C. J. Lavie, and A. C. Dornelles, "Impact of Coenzyme Q-10 on Parameters of Cardiorespiratory Fitness and Muscle Performance in Older Athletes Taking Statins," *Physician and Sportsmedicine* 40, no. 4 (2012): 88–95, doi: 10.3810/psm.2012.11.1991.

217 **fish oil supplements cause cancer:** T. M. Brasky, A. K. Darke, X. Song, C. M. Tangen, P. J. Goodman, I. M. Thompson, F. L. Meyskens Jr., et al., "Plasma Phospholipid Fatty Acids and Prostate Cancer Risk in the SELECT Trial," *Journal of the National Cancer Institute* 105, no. 15 (2013): 1132–41, doi: 10.1093/jnci/djt174.

217 **In a paper that we quickly published in another journal:** J. J. DiNicolantonio, M. F. McCarty, C. J. Lavie, and J. H. O'Keefe, "Do Omega-3 Fatty Acids Cause Prostate Cancer?" *Missouri Medicine* 110, no. 4 (2013): 293–5. For our response to the fish oil study, see: J. J. DiNicolantonio, M. F. McCarty, C. J. Lavie, and J. H. O'Keefe, "Scientific Correspondence: Do Omega-3 Fatty Acids Cause Prostate Cancer?" *Missouri Medicine*, July–August 2013, http://www.omagdigital.com/article/SCIENTIFIC+CORRESPONDENCE/1476759/0/article.html.

218 **Vitamin D deficiency is not just about an increased risk for weak, soft bones:** For a thorough discussion of the importance of vitamin D, see: Holick, *The Vitamin D Solution*.

220 **It's one of the oldest remedies known to humankind:** For a quick review of aspirin's history, see: Elizabeth Landau, "From a Tree, a 'Miracle' Called Aspirin," CNN.com, December 22, 2010, http://www.cnn.com/2010/HEALTH/12/22/aspirin.history/.

224 **And in a surprising twist of conventional wisdom:** Horwich and Fonarow, "Reverse Epidemiology Beyond Dialysis Patients."

Epilogue

226 **Semmelweis reflex:** "Dr. Semmelweis' Biography," Semmelweis Society International, accessed October 22, 2013, http://semmelweis.org/about/dr-semmelweis-biography/.

226 **Puerperal fever was a harrowing disease:** C. Hallett, "The Attempt to Understand Puerperal Fever in the Eighteenth and Early Nineteenth Centuries: The Influence of Inflammation Theory," *Medical History* 49, no. 1 (2005): 1–28.

231 **the power of the human microbiome:** To follow all the latest news and studies on the human microbiome, see the dedicated page on the National Institute of Health's website: http://commonfund.nih.gov/hmp/ (accessed October 22, 2013).

231 **as reflected in the arts and literature:** Eknoyan, "A History of Obesity."

232 **YOUNG ADULTS ONLY WALK FIVE MINUTES A DAY "BECAUSE OF TECHNOLOGY":** Radhika Sanghani, "Young Adults Only Walk Five Minutes a Day 'Because of Technology,' " *Telegraph*, October 19, 2013, http://www.telegraph.co.uk/technology/news/10389466/Young-adults-only-walk-five-minutes-a-day-because-of-technology.html.

233 **"survival of the fittest":** *Encyclopaedia Britannica*, s. v. "survival of the fittest," accessed October 22, 2013, http://www.britannica.com/EBchecked/topic/575460/survival-of-the-fittest.

Index